Grade 5

Reveal MATH®

Student Practice Book

McGraw Hill

mheducation.com/prek-12

Copyright © 2022 McGraw Hill

All rights reserved. The contents, or parts thereof, may be reproduced in print form for non-profit educational use with *Reveal Math*, provided such reproductions bear copyright notice, but may not be reproduced in any form for any other purpose without the prior written consent of McGraw Hill, including, but not limited to, network storage or transmission, or broadcast for distance learning.

Send all inquiries to:
McGraw Hill
8787 Orion Place
Columbus, OH 43240

ISBN: 978-0-07-693710-3
MHID: 0-07-693710-0

Printed in the United States of America.

7 8 9 LHS 25 24 23 22

Grade 5
Table of Contents

Unit 2
Volume
Lessons
- 2-1 Understand Volume...1
- 2-2 Use Unit Cubes to Determine Volume 3
- 2-3 Use Formulas to Determine Volume..............................5
- 2-4 Determine Volume of Composite Figures 7
- 2-5 Solve Problems Involving Volume................................9

Unit 3
Place Value and Number Relationships
Lessons
- 3-1 Generalize Place Value..11
- 3-2 Extend Place Value to Decimals 13
- 3-3 Read and Write Decimals 15
- 3-4 Compare Decimals...17
- 3-5 Use Place Value to Round Decimals.............................19

Unit 4
Add and Subtract Decimals
Lessons
- 4-1 Estimate Sums and Differences of Decimals....................21
- 4-2 Represent Addition of Decimals 23
- 4-3 Represent Addition of Tenths and Hundredths..................25
- 4-4 Use Partial Sums to Add Decimals 27
- 4-5 Represent Subtraction of Decimals..............................29
- 4-6 Represent Subtraction of Tenths and Hundredths...............31
- 4-7 Strategies to Subtract Decimals 33
- 4-8 Explain Strategies to Add and Subtract Decimals 35

Unit 5
Multiply Multi-Digit Whole Numbers

Lessons

- 5-1 Understand Powers and Exponents 37
- 5-2 Patterns When Multiplying a Whole Number by Powers of 10 39
- 5-3 Estimate Products of Multi-Digit Factors 41
- 5-4 Use Area Models to Multiply Multi-Digit Factors 43
- 5-5 Use Partial Products to Multiply Multi-Digit Factors 45
- 5-6 Relate Partial Products to an Algorithm 47
- 5-7 Multiply Multi-Digit Factors Fluently 49

Unit 6
Multiply Decimals

Lessons

- 6-1 Patterns When Multiplying Decimals by Powers of 10 51
- 6-2 Estimate Products of Decimals 53
- 6-3 Represent Multiplication of Decimals 55
- 6-4 Use an Area Model to Multiply Decimals 57
- 6-5 Generalizations about Multiplying Decimals 59
- 6-6 Explain Strategies to Multiply Decimals 61

Unit 7
Divide Whole Numbers

Lessons

- 7-1 Division Patterns with Multi-Digit Numbers 63
- 7-2 Estimate Quotients 65
- 7-3 Relate Multiplication and Division of Multi-Digit Numbers 67
- 7-4 Represent Division of 2-Digit Divisors 69
- 7-5 Use Partial Quotients to Divide 71
- 7-6 Divide Multi-Digit Whole Numbers 73
- 7-7 Solve Problems Involving Division 75

Unit 8
Divide Decimals
Lessons
- 8-1 Division Patterns with Decimals and Powers of 10 77
- 8-2 Estimate Quotients of Decimals 79
- 8-3 Represent Division of Decimals by a Whole Number 81
- 8-4 Divide Decimals by Whole Numbers 83
- 8-5 Divide Whole Numbers by Decimals 85
- 8-6 Divide Decimals by Decimals 87

Unit 9
Add and Subtract Fractions
Lessons
- 9-1 Estimating Sums and Differences of Fractions 89
- 9-2 Represent Addition of Fractions with Unlike Denominators 91
- 9-3 Add Fractions with Unlike Denominators 93
- 9-4 Represent Subtraction of Fractions with Unlike Denominators 95
- 9-5 Subtract Fractions with Unlike Denominators 97
- 9-6 Add Mixed Numbers with Unlike Denominators 99
- 9-7 Subtract Mixed Numbers with Unlike Denominators 101
- 9-8 Add and Subtract Mixed Numbers with Regrouping 103
- 9-9 Solve Problems Involving Fractions and Mixed Numbers 105

Unit 10
Multiply Fractions
Lessons
- 10-1 Represent Multiplication of a Whole Number by a Fraction 107
- 10-2 Multiply a Whole Number by a Fraction 109
- 10-3 Represent Multiplication of a Fraction by a Fraction 111
- 10-4 Multiply a Fraction by a Fraction 113
- 10-5 Determine the Area of Rectangles with Fractional Side Lengths 115
- 10-6 Represent Multiplication of Mixed Numbers 117
- 10-7 Multiply Mixed Numbers 119
- 10-8 Multiplication as Scaling 121
- 10-9 Solve Problems Involving Fractions 123

Unit 11
Divide Fractions

Lessons

- 11-1 Relate Fractions to Division 125
- 11-2 Solve Problems Involving Division 127
- 11-3 Represent Division of Whole Numbers by Unit Fractions 129
- 11-4 Divide Whole Numbers by Unit Fractions 131
- 11-5 Represent Division of Unit Fractions by Non-Zero Whole Numbers 133
- 11-6 Divide Unit Fractions by Non-Zero Whole Numbers 135
- 11-7 Solve Problems Involving Fractions 137

Unit 12
Measurement and Data

Lessons

- 12-1 Convert Customary Units 139
- 12-2 Convert Metric Units 141
- 12-3 Solve Multi-Step Problems Involving Measurement Units 143
- 12-4 Represent Measurement Data on a Line Plot 145
- 12-5 Solve Problems Involving Measurement Data on Line Plots ... 147

Unit 13
Geometry

Lessons

- 13-1 Understand the Coordinate Plane 149
- 13-2 Plot Ordered Pairs on the Coordinate Plane 151
- 13-3 Represent Problems on a Coordinate Plane 153
- 13-4 Classify Triangles by Properties 155
- 13-5 Properties of Quadrilaterals 157
- 13-6 Classify Quadrilaterals by Properties 159

Unit 14
Algebraic Thinking

Lessons

- **14-1** Write Numerical Expressions 161
- **14-2** Interpret Numerical Expressions 163
- **14-3** Evaluate Numerical Expressions 165
- **14-4** Numerical Patterns 167
- **14-5** Relate Numerical Patterns 169
- **14-6** Graphs of Numerical Patterns 171

Lesson 2-3
Additional Practice

Name _Sheyla Rodriguez Casillas_

> **Review**
>
> **You can calculate the volume of a rectangular prism by multiplying its length, width, and height.**
>
> A rectangular prism has a length of 6 centimeters, a width of 2 centimeters, and a height of 3 centimeters.
> $V = l \times w \times h = 6 \times 2 \times 3 = 36$ cubic cm.
> $V = B \times h = 12 \times 3 = 36$ cubic cm.

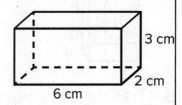

Use the given volume formula to calculate the volume of each rectangular prism. $V = l \times w \times h$

1. $V = \underline{\ 5\ }$ in. $\times \underline{\ 4\ }$ in. $\times \underline{\ 3\ }$ in.

    $V = \underline{\ 60\ }$ cubic in.

    $\begin{array}{r} 20 \\ \times\ 3 \\ \hline 60 \end{array}$

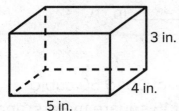

2. $V = \underline{\ 7\ }$ cm $\times \underline{\ 3\ }$ cm $\times \underline{\ 4\ }$ cm

    $V = \underline{\ 84\ }$ cubic cm

    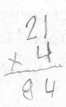

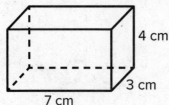

Student Practice Book

Label the dimensions of each rectangular prism. Then use the volume formula to calculate the volume of each prism. $V = B \times h$

3. $V = \underline{9} \times \underline{6}$
   $V = \underline{54}$ cubic units

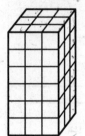

4. $V = \underline{24} \times \underline{5}$
   $V = \underline{120}$ cubic units

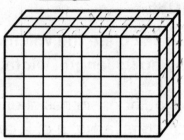

5. Use the two volume formulas, $V = l \times w \times h$ and $V = B \times h$, to write multiplication equations for the volume of the rectangular prism.

   $V = \underline{6} \times \underline{4} \times \underline{2}$
   $V = \underline{24} \times \underline{2}$

6. Calculate the volume of a video game console with the dimensions 9 inches by 11 inches by 3 inches.

   $V = \underline{9}$ in. $\times \underline{11}$ in. $\times \underline{3}$ in.
   $V = \underline{27}$ cubic in.

7. A window air conditioner can cool a space of up to 50 cubic meters. The floor of a room has an area of 16 square meters, and the height of the walls is 3 meters. Will the air conditioner be able to cool the room? Explain.

Math @ Home Activity

Use a ruler or tape measure to measure the dimensions of boxes. Round dimensions to the nearest inch or centimeter. Ask your child to use the volume formula $V = l \times w \times h$ to find the volume of each box.

Student Practice Book

Lesson 2-4
Additional Practice

Name _Sheyla Rodriguez Casillas_

Review

You can find the volume of a composite solid figure by decomposing the figure, finding the volume of each solid figure, and then adding to find the total volume.

The composite solid figure can be decomposed into rectangular prisms with volumes of 36 cubic units and 48 cubic units. The total volume of the composite solid figure is $36 + 48 = 84$ cubic units.

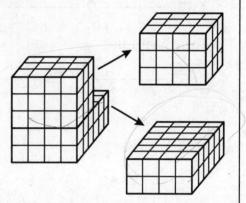

Find the volume of each composite solid figure.

1. $V =$ __256__ cubic units

2. $V =$ _____ cubic units

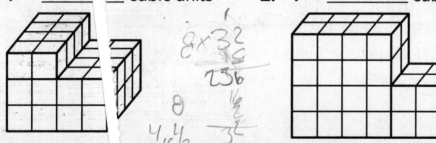

Find the volume of each figure. Draw line(s) to show how you decomposed the figure.

3. $V =$ _____ cubic units

4. $V =$ _____ cubic units

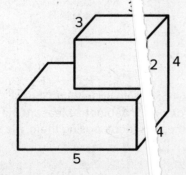
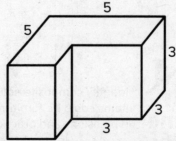

5. Find the volume of concrete used to create these steps. Explain your work.

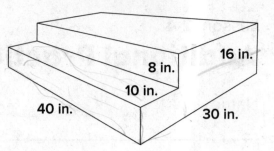

Top Prism:
V = _____ × _____ × _____ = _____ cubic in.

Bottom Prism:
V = _____ × _____ × _____ = _____ cubic in.

Total Volume:
V = _____ + _____ = _____ cubic in.

6. The shape of an entertainment center is shown. Calculate its volume. Explain your work.

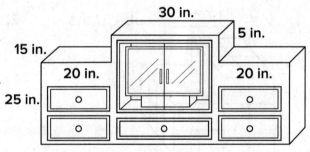

Identify composite solid figures around the home with your child. These items could be furniture such as dressers, desks, or cabinets. Measure the dimensions and practice finding their volumes by decomposing them.

Student Practice Book

Lesson 2-5
Additional Practice

Name

Review

You can solve problems involving volume by using the formulas $V = l \times w \times h$ and $V = B \times h$.

The unknown length of the rectangular prism can be found by substituting the values for the volume, width, and height into the volume formula.

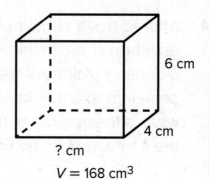

$$V = l \times w \times h$$
$$168 = l \times 4 \times 6$$
$$168 = l \times 24$$
$$168 \div 24 = l$$
$$7 = l$$

The unknown length is 7 centimeters.

Solve.

1. The volume of the dresser is 24 cubic feet. How tall is the dresser? Explain.

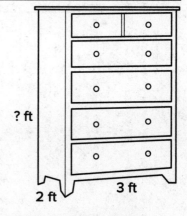

2. The volume of the rectangular prism is 432 cubic inches. What is the width of the prism?

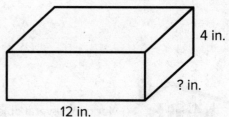

3. A freezer has a volume of 54 cubic feet. It has a length of 6 feet and a height of 3 feet. What is the width of the freezer? Explain.

4. Andrea has a container in the shape of a rectangular prism that she uses for blueberry picking. If the blueberries fill the container to a height of 15 centimeters, what is the volume of the blueberries in the container? Show your work.

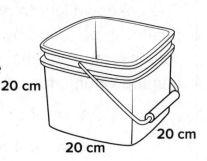

5. A kitchen island has a volume of 42 cubic feet. Use the dimensions given in the figure to find the missing dimension. Explain.

Identify several objects at home that can be represented as rectangular prisms. Have your child predict which object has the greatest volume, and then calculate the volumes of the objects to confirm or refute the prediction.

Lesson 3-3
Additional Practice

Name _____

> ### Review
>
> **You can write decimals to the thousandths using standard form, expanded form, and word form.**
>
> | tens | ones | tenths | hundredths | thousandths |
> |:---:|:---:|:---:|:---:|:---:|
> | 2 | 7 | 5 | 4 | 9 |
>
> In the chart, 27.549 is written in standard form. Write the number in expanded form and word form.
>
> Remember to write the decimal point as "and" when writing the number in word form.
>
> twenty-seven and five hundred forty-nine thousandths
>
> When writing the number in expanded form, multiply each digit by its place value in decimal form.
>
> (2 × 10) + (7 × 1) + (5 × 0.1) + (4 × 0.01) + (9 × 0.001)

1. A piece of ribbon is 3.75 feet long. Write 3.75 in expanded form using fractions.

    (3×1) + (7×0.1) + (5×0.01)

2. Write 59.107 in expanded form. Use the place-value chart to find the value of each digit.

tens	ones	tenths	hundredths	thousandths
5	9	1	0	7

(5×10) + (9×1) + (1×0.1) + (0×0.01)

(7×0.007) + _____ + _____ + _____

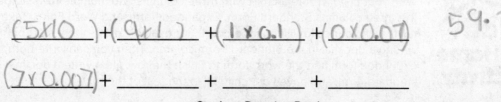

Student Practice Book
15

3. Write each decimal in word form.

   a. 13.5 = ~~thourt a three drsmal fave~~

   b. 1.35 = ~~one and tcu~~

   c. 0.135 = _____

   d. 0.013 = _____

4. Write each decimal in standard form.

   a. two and nine tenths = _____

   b. twenty-nine and six hundredths = _____

   c. six and twenty-five thousandths = _____

   d. eight hundred forty-one thousandths = _____

5. Write the standard form of each number written in expanded form.

   a. $3 + \frac{8}{10} + \frac{2}{1,000} =$ _____

   b. $30 + 8 + \frac{9}{100} =$ _____

   c. $70 + 0.08 + 0.002 =$ _____

   d. $1 + 0.5 + 0.09 =$ _____

6. Colby says that $\frac{27}{100}$ written in word form is twenty-seven thousandths. Do you agree? Explain.

With your child, create a chart with three columns and multiple rows. Label the three columns Standard Form, Expanded Form, and Word Form. Take turns with your child filling in the chart. For example, have them start by writing a decimal in the Standard Form column. Then you can write both the expanded form and the word form to finish the row. Next write a decimal in standard form, and have your child fill in the rest of the row.

Student Practice Book

Lesson 3-4

Additional Practice

Name _____

Review

You can compare two decimals to the thousandths place.

Lulu runs a mile in 9.375 minutes, and Kindra runs a mile in 9.376 minutes. Compare the two decimals.

Line up the numbers on the decimal point so all the place values will be lined up. Both numbers have 9 ones, 3 tenths, and 7 hundredths.

Since the digits in the thousandths place are different, compare those two digits.

9.375
9.376

9.375 < 9.376

---

1. Efren has 0.3 ounce of water and 0.38 ounce of salt. Line up the numbers on the decimal point to determine which amount is less than the other amount.

   0.3
   0.38

   0.3 ounce ◯ 0.38 ounce

2. Write thirty-seven and forty-nine hundredths in standard form. _____

   Is the number greater than or less than 37.45?

   thirty-seven and forty-nine hundredths ◯ 37.45

Student Practice Book

3. Write a decimal that is equal to 1.5. Explain your answer.

4. Which of the following are correct? Choose all that apply.

   A. 0.09 > 0.009

   B. 1.26 < 1.258

   C. 29.99 = 29.990

   D. 37.48 > 37.461

   E. 5.908 = 5.980

5. Lorinda has $10.81 in her piggy bank. Thi has $10.18 in his piggy bank. Compare the amounts.

   $10.81 ◯ $10.18

6. Lincoln bikes 24.28 miles on Monday and 24.385 miles on Tuesday. Compare the distances.

   24.28 ◯ 24.385

7. Jewel and Karl are playing a game. Jewel has 15.42 points. Karl has 15.428 points. Compare the number of points. Who has the greater number of points?

   15.42 ◯ 15.428

   _____ has the greater number of points.

8. Zina is 4.25 feet tall. Her cousin Sam is 4.175 feet tall. Compare the heights. Who is taller?

   4.25 ◯ 4.175

   _____ is taller.

Give your child three index cards, and have them write >, <, and = on the cards. Identify numbers around your home that are written in decimal form. Ask your child to hold up the correct card to compare the numbers. The cards can also be used at a park to compare lengths of trails, at a gas station to compare prices of gasoline, or at a grocery store to compare prices of food items.

Student Practice Book

Lesson 3-5
Additional Practice

Name _____

Review

You can round decimals.

Marg has 14.875 feet of rope. Round the length of the rope to the nearest tenth.

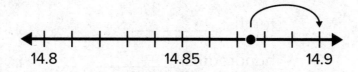

14.875 feet rounded to the nearest tenth is 14.9 feet.

1. Round each decimal to the nearest whole number.

   a. 0.948 _____   b. 34.972 _____

   c. 4.013 _____   d. 48.671 _____

   e. 9.05 _____   f. 56.143 _____

   g. 12.489 _____   h. 66.701 _____

   i. 20.87 _____   j. 79.862 _____

   k. 26.187 _____   l. 92.557 _____

2. Round each decimal to the given place value.

   a. Round 1.521 to the nearest tenth. _____

   b. Round 4.037 to the nearest hundredth. _____

   c. Round 19.232 to the nearest tenth. _____

   d. Round 41.691 to the nearest hundredth. _____

   e. Round 83.888 to the nearest tenth. _____

3. Round each decimal to the given place.

   a. 0.143   tenths: _____

   hundredths: _____

   b. 10.976   ones: _____

   hundredths: _____

   c. 39.183   ones: _____

   tenths: _____

   d. 71.565   tenths: _____

   hundredths: _____

4. A puppy weighs 10.49 pounds. To the nearest tenth, *about* how much does the puppy weigh?

   _____ pounds

5. Dona says that 35.284 rounded to the nearest hundredth is 35.29. Do you agree? Explain.

6. There is $78.69 in a checking account. The amount needs to be rounded to the nearest whole dollar. Wally says there is about $78 in the account, and Tu says there is about $79 in the account. Who is correct? Explain.

Discuss instances where rounding can be helpful around your home. For example, if you need 0.75 gallon of milk, you can round up and buy 1 gallon of milk. Then ask your child to round different decimal values. Practice rounding the values to the nearest whole number, tenth, and hundredth.

Student Practice Book

Lesson 4–3
Additional Practice

Name

Review

You can add tenths and hundredths using decimal grids.

The distance between a school and a playground is 0.8 mile. There is a parking lot farther down the road. The distance between the playground and the parking lot is 0.45 mile. What is the distance between the school and the parking lot?

You can represent the problem with the equation $0.8 + 0.45 = d$. Think of the number of tenths as a number of hundredths: 8 tenths = 80 hundredths, and $0.8 = 0.80$. Use decimal grids to add.

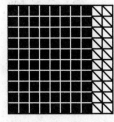
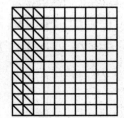

The distance between the school and the parking lot is 1.25 miles.

Use decimal grids to find each sum.

1. $0.7 + 0.09 = $ _____

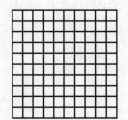
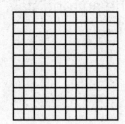

2. $0.32 + 0.5 = $ _____

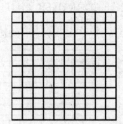
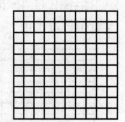

Student Practice Book
25

Use decimal grids to solve.

3. 0.2 + 0.48 = _____   **4.** 0.35 + 0.9 = _____

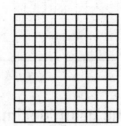 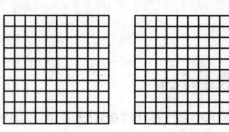

5. A frog jumped 0.7 meter and then jumped 0.65 meter. What is the total distance that the frog jumped?

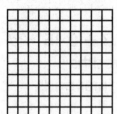 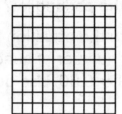

6. Dean is buying school supplies at the bookstore. Pencils cost $0.75 each and erasers cost $0.50 each. What is the total if he buys one pencil and one eraser?

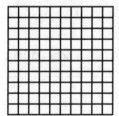 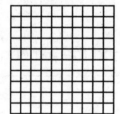

7. Donia drank 0.6 liter of water during the first half of the game. She drank 0.45 liter of water during the second half. How much water did Donia drink during the game?

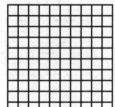 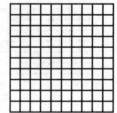

Math @ Home Activity

Use a group of dimes and a group of nickels and pennies to represent adding tenths and hundredths. Ask your child to write the amount for each group and then use the decimal grids to add the amounts. Then have him or her use decimal grids to add the amounts. Have your child write the total amount of money. Then repeat with different numbers of coins in each group.

Student Practice Book

Lesson 4-4

Additional Practice

Name

Review

You can decompose numbers to add decimals.

Franklin has 16.25 pounds of apples and 12.05 pounds of oranges. Find the total weight of the fruit Franklin has.

The equation 16.25 + 12.05 = f represents the situation. Decompose both addends by place value or by whole numbers and decimals. Then add the partial sums.

By place value

10 + 10 = 20
6 + 2 = 8
0.2 + 0.0 = 0.2
0.05 + 0.05 = 0.1
20 + 8 + 0.2 + 0.1 = 28.3

By whole numbers and decimals

16 + 12 = 28   0.25 + 0.05 = 0.30
28 + 0.3 = 28.3

Franklin has 28.3 pounds of fruit.

Use partial sums to add.

1. 3.16 + 8.4

2. 17.85 + 0.5

3. 25.42 + 16.71

4. 70.94 + 59.01

Student Practice Book

Solve.

5. A city had 6.95 inches of snow fall on Tuesday and 8.25 inches of snow fall on Wednesday. How much snow fell over the two-day period?

6. A plant grew 7.3 centimeters one month and 12.15 centimeters the next month. By how many centimeters did the plant grow over the two-month period?

7. On a bike trip, the group rode 48.52 miles the first day and 57.6 miles the second day. How far did the group ride during the first two days of the trip?

8. Jeremy wants to save up $10 to buy a poster. He earned $6.32 last week by collecting and selling aluminum cans. He earned $3.58 this week. Has Jeremy earned enough money to buy the poster? Explain.

Math @ Home Activity

Look for situations around your home where it would be natural to add decimals. Have your child decompose the addends to find the sum and explain their method. Have him or her explain their method. For example, if the sum of the prices of two items needs to be found, have them decompose the amounts and tell whether they used place values or whole numbers and decimals.

Student Practice Book

Lesson 4-5
Additional Practice

Name

> **Review**
>
> **You can represent subtraction of decimals using a number line or a decimal grid.**
>
> Rina has $0.90, and Joel has $0.70. How much more money does Rina have than Joel?
>
> You can represent this problem with the equation $0.9 - 0.7 = d$. Use a number line or a decimal grid to find the difference.
>
>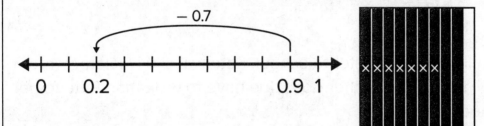
>
> The difference is 0.2. So, Rina has $0.20 more than Joel.

Use a number line to solve each equation.

1. $0.6 - 0.5 = $ _____

2. $0.05 - 0.01 = $ _____

Student Practice Book
29

Use a decimal grid to solve each equation.

3. 0.31 − 0.24 = _____

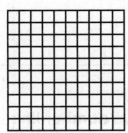

4. 0.8 − 0.4 = _____

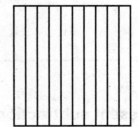

Solve.

5. Janelle has a length of string that is 0.8 meter long. She cuts off a piece that is 0.5 meter long. How long is the piece of string that remains?

6. Joe walks 0.42 kilometer to school. Pete walks 0.17 kilometer to school. How much farther does Joe have to walk than Pete to get to school?

7. On Friday, 0.7 inch of rain fell. On Saturday, the amount of rain that fell was 0.2 inch less than the amount that fell on Friday. How much rain fell on Saturday?

8. Avelina has 0.83 GB of space left on her memory card. She adds photos that take up an additional 0.24 GB of space. How much space is left on the memory card?

With your child, practice representing subtraction with decimals using coins. Show your child a group of coins and then model taking some of the coins away. Have your child write an equation to model the problem, and then use a number line or a decimal grid to find out how much money is left. Then have your child count the remaining money to verify their answers.

Student Practice Book
30

Lesson 4–6
Additional Practice

Name _____

> **Review**
>
> **You can subtract decimals that have different place values.**
>
> Emma walks 0.9 mile on Saturday and 0.62 mile on Sunday. How much farther does Emma walk on Saturday than on Sunday?
>
> You can represent the problem with the equation $0.9 - 0.62 = d$. Think of the number of tenths as a number of hundredths: 9 tenths = 90 hundredths, and 0.9 = 0.90. Use decimal grids to subtract.
>
>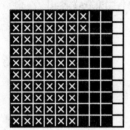
>
> Emma walks 0.28 mile more on Saturday than Sunday.

Use decimal grids to solve each equation.

1. $0.5 - 0.04 =$ _____

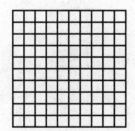

2. $0.66 - 0.3 =$ _____

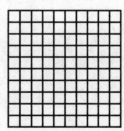

3. $1.17 - 0.6 =$ _____

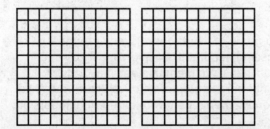

4. $1.9 - 0.58 =$ _____

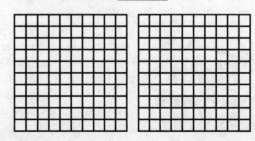

Student Practice Book

Solve.

5. A basketball player brings 1.8 liters of water to practice. If she has 0.65 liter of water left after practice, how much water did the player drink during practice?

6. Suzie buys 1.4 pounds of Swiss cheese and 0.72 pound of cheddar cheese. How much more Swiss cheese did Suzie buy?

7. Yesterday, 0.63 inch of snow fell. Today, 0.2 inch of snow fell. How much more snow fell yesterday than today?

8. Charles bought 1.28 pounds of peanuts. He ate 0.4 pound of the peanuts at the game. What is the weight of the peanuts that Charles still has left?

Use a group of dimes and a group of dimes, nickels, and pennies to represent subtracting tenths and hundredths. Have your child write the amounts for each group. Then have them use decimal grids to subtract the lesser amount from the greater amount. Have your child write the total amount of money. Then repeat with different numbers of coins in each group.

Student Practice Book

Lesson 4-7

Additional Practice

Name _____

> ### Review
>
> **You can use strategies to subtract decimals.**
>
> Charla has $21.74, and Clair has $13.20. How much more money does Charla have than Clair?
>
> To solve, find $21.74 − $13.20.
>
> | Use partial differences. Decompose 13.20 by place value and subtract. | Use counting on. |
> |---|---|
> | 21.74 − 10 = 11.74 | 13.20 + 7 = 20.20 |
> | 11.74 − 3 = 8.74 | 20.20 + 0.8 = 21 |
> | 8.74 − 0.2 = 8.54 | 21 + 0.74 = 21.74 |
> | | 7 + 0.8 + 0.74 = 8.54 |
>
> Charla has $8.54 more than Clair.

Find each difference. Show your work.

1. 4.74 − 1.9 = _____

2. 8.8 − 0.67 = _____

3. 41.53 − 17.36 = _____

4. 55.91 − 16.9 = _____

Student Practice Book

Find each difference. Explain your method.

5. 34.92 − 15.75 = _____

6. 23.16 − 12.84 = _____

Solve.

7. A box of books weighs 42.38 pounds. After taking out some of the books, the box now weighs 25.75 pounds. What is the weight of the books that were taken out of the box?

8. Noelle has hiked 1.38 kilometers along the trail from the nature center. The waterfall is 3.2 kilometers from the nature center. How much farther does Noelle have left to hike to get to the waterfall?

Help your child practice decomposing numbers in order to subtract. Write several subtraction equations on the left side of a sheet of paper. In a mixed-up order, write the decomposed place values on the right side of the paper. Have your child find the matching equation and decomposition. Then work with your child to solve each equation.

Student Practice Book

Lesson 4-8
Additional Practice

Name _____

> **Review**
>
> **You can use any method to add or subtract decimals.**
>
> Federico rode his bicycle 3.8 kilometers today. His coach wants him to increase the distance by 0.75 kilometer each day. What distance will Federico ride his bicycle tomorrow?
>
> You can represent this problem with a bar diagram. The bar diagram shows that you have to solve the equation $3.8 + 0.75 = d$.
>
> One way to solve is to decompose the numbers to add.
>
> $3.8 = 3 + 0.8 + 0.00$ and $0.75 = 0 + 0.7 + 0.05$
>
> $3 + 0 = 3 \quad 0.8 + 0.7 = 1.5 \quad 0.00 + 0.05 = 0.05$
>
> Add the partial sums: $3 + 1.5 + 0.05 = 4.55$
>
> Federico will ride 4.55 kilometers tomorrow.

Solve each equation. Explain how you determined which strategy to use.

1. $5.83 + 2.8 =$ _____

2. $9.4 - 6.58 =$ _____

Student Practice Book

Solve. Explain the strategy used to solve.

3. A dog drinks 2.2 liters of water each day and a cat drinks 0.86 liter of water each day. How many liters of water do the dog and cat drink in all?

4. The length of one earthworm is 6.7 centimeters. The length of a second earthworm is 5.47 centimeters. How much longer is the first earthworm than the second earthworm?

5. A gardener ordered 13.25 kilograms of gravel for a project. The gardener had to order an additional 11.9 kilograms of gravel to complete the project. How much gravel was needed?

6. Truman buys 34.7 meters of fencing to enclose his pond. He uses only 29.56 meters of the fencing. How much fencing is left over?

Give your child a ruler, and have them measure the lengths of two different objects in centimeters. Then have them create a bar diagram that represents finding the difference between the measurements, and then write and solve an equation. Then repeat the activity, this time having your child find the sum of the lengths of the two objects.

Student Practice Book

Lesson 5-1
Additional Practice

Name _____

> **Review**
>
> **You can use a base and an exponent to find and represent powers of 10.**
>
> Find 10^5.
>
> In the expression 10^5, 10 is the *base* and 5 is the *exponent*. Exponents represent how many times the base is used as a factor.
>
> $$10^5 = 10 \times 10 \times 10 \times 10 \times 10 = 100{,}000$$
>
> Represent $10 \times 10 \times 10$ in exponential form, using a base and an exponent.
>
> Since 10 is the repeated factor, and it is being used as a factor 3 times, 10 will be the base and 3 will be the exponent. Write the expression in exponential form.
>
> $$10 \times 10 \times 10 = 10^3$$

Write each power of 10 as a product of 10s.

1. 10^2
2. 10^4

3. 10^6
4. 10^3

Write the exponential form.

5. $10 \times 10 \times 10 =$ _____
6. $10 \times 10 \times 10 \times 10 \times 10 =$ _____

7. $10 \times 10 \times 10 \times 10 =$ _____
8. $10 \times 10 =$ _____

Write the exponential form of each power of 10.

9. 10 = _____

10. 10,000 = _____

11. 100 = _____

12. 10,000,000 = _____

13. A company has a total of 10^5 employees. How many employees does the company have?

The company has _____ employees.

14. Aletha has 10^3 photos on her computer. How many photos does she have?

Aletha has _____ photos.

15. Desmond drives 100 miles each day. What is this number as a product of 10s? What is this number in exponential form?

16. Each day, Maurice has a goal to walk 10,000 steps. What is this number as a product of 10s? What is this number in exponential form?

Create three sets of six note cards. One set should include the powers of 10 in exponential form from 10^1 to 10^6. The second set should include the powers of 10 written in standard form, from 10 to 1,000,000. The third set should include each number written as a product of 10s, from 10 to 10 × 10 × 10 × 10 × 10 × 10. Mix the cards and arrange them face down in an array of the 18 cards. Play a matching game, in which each player gets to pick three cards, trying to match the three equivalent forms of each number. Play until all the cards are matched.

Student Practice Book

Lesson **5-2**

Additional Practice

Name _____

> **Review**
>
> **You can use a pattern to multiply whole numbers by powers of 10.**
>
> A company is purchasing pens. There are 16 pens in a pack, and the company purchases 1,000 packs. How many pens does the company purchase?
>
> When multiplying a number by a power of 10, the product becomes 10 times greater for each power of 10.
>
> $16 \times 10^1 = 16 \times 10 = 160$
>
> $16 \times 10^2 = 16 \times 100 = 1,600$
>
> $16 \times 10^3 = 16 \times 1,000 = 16,000$
>
> The company purchases 16,000 pens.

Use patterns to find each product.

1. $25 \times 10 =$ _____

    $25 \times 100 =$ _____

    $25 \times 1,000 =$ _____

2. $73 \times 1,000 =$ _____

    $73 \times 10,000 =$ _____

    $73 \times 100,000 =$ _____

3. $44 \times 10^2 =$ _____

    $44 \times 10^3 =$ _____

    $44 \times 10^4 =$ _____

4. $81 \times 10^4 =$ _____

    $81 \times 10^5 =$ _____

    $81 \times 10^6 =$ _____

Student Practice Book

Use patterns to find the value of each expression.

5. $69 \times 10^3 = $ _____

6. $247 \times 10^2 = $ _____

7. $80 \times 10^4 = $ _____

8. $506 \times 10^1 = $ _____

Complete each equation with a power of 10.

9. $23 \times $ _____ $= 2{,}300$

10. $71 \times $ _____ $= 7{,}100{,}000$

11. $9 \times $ _____ $= 9{,}000$

12. $18 \times $ _____ $= 18{,}000{,}000$

13. Hershel thinks that $30 \times 1{,}000 = 30{,}000$. How would you respond to Hershel?

14. Which equations are true? Choose all that apply.

   A. $8 \times 10 = 80$

   B. $70 \times 100 = 7{,}000$

   C. $29 \times 1{,}000 = 2{,}900$

   D. $60 \times 100 = 6{,}000$

Use multiple resources to find numbers in context that are whole numbers multiplied by a power of 10. Have your child copy the number from the source and then write the number as a product of a whole number and a power of 10. For example, if 28,000 people attended the game, this could be expressed as $28 \times 1{,}000$.

Student Practice Book

Lesson **5-3**
Additional Practice

Name _____

> **Review**
>
> **You can use rounding or compatible numbers to estimate a product.**
>
> There are 329 students in a grade school. Each student donates 11 canned goods. About how many canned goods does the school collect?
>
> 329 × 11 = c   329 rounds to 330.
>   ↓   ↓   11 rounds to 10.
> 330 × 10 = 3,300
>
> A reasonable estimate is that the school collected 3,300 canned goods.

Estimate each product.

1. 412 × 17 = _____ × _____ = _____

2. 281 × 32 = _____ × _____ = _____

3. 81 × 687 = _____ × _____ = _____

4. 57 × 509 = _____ × _____ = _____

5. 749 × 64 = _____ × _____ = _____

6. 499 × 51 = _____ × _____ = _____

7. 79 × 643 = _____ × _____ = _____

8. 24 × 702 = _____ × _____ = _____

Student Practice Book

9. One sleeve of plastic cups holds 32 cups. A store has 73 sleeves of cups on hand. About how many plastic cups does the store have on hand? Show your work.

10. One student ticket for the zoo costs $12. If a class of 89 students goes to the zoo, about how much will the tickets cost? Show your work.

Determine whether the calculation is reasonable. Explain why or why not.

11. Patti multiples 713 by 58 to get 41,354. Is her calculation reasonable?

12. Raul sells 321 sports passes for $14 each. He says that he collects $494. Is his calculation reasonable?

Use situations around your home to help your child practice estimating. For example, when selling fundraising items or buying items in bulk, ask your child to estimate the total cost. In some situations, tell your child the total cost and ask them to use estimation to determine whether it is reasonable.

Lesson 5-4
Additional Practice

Name _____

> **Review**
>
> **You can use area models and partial products to decompose factors and find products.**
>
> A parking lot is 248 feet long and 56 feet wide. What is the area of the parking lot?
>
> To find the area, solve 56 × 248. Use an area model.
>
>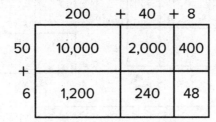
>
> Add the partial products to find the product.
>
> 10,000 + 2,000 + 400 + 1,200 + 240 + 48 = 13,888
>
> The area of the parking lot is 13,888 square feet.

Use the area model and partial products to solve.

1. Jody's house has a rectangular deck attached. What is the area of the deck?

   The area of the deck is _____ square yards.

Student Practice Book

Use area models and partial products to solve.

2. 74 × 49 = _____   **3.** 18 × 221 = _____

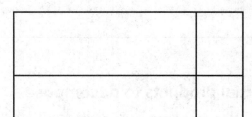

4. 65 × 734 = _____   **5.** 32 × 603 = _____

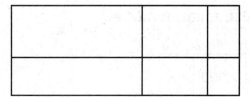

6. What is the area of the rectangular field?

168 ft

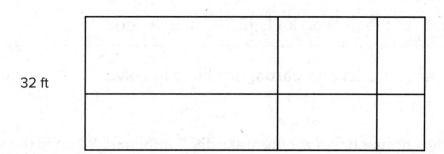

32 ft

The area of the field is _____ square feet.

7. A sporting goods company ships out 621 boxes of baseballs each day. How many boxes of baseballs are shipped out in 25 days?

The company ships out _____ boxes of baseballs.

Measure the length and width of a rectangular room in your home. Use string, yarn, books, rulers, or painter's tape, etc., to make the room look like an area model, or just draw the room on a piece of paper. Then have them find the area of the room using an area model and partial products. Have your child explain how he or she calculated the area.

Student Practice Book

Lesson **5-5**

Additional Practice

Name _____

Review

You can use partial products to find products of multi-digit factors.

Find 16 × 128. Decompose the factors by place value. Then use partial products.

$$\begin{array}{r} 128 \\ \times\ 16 \\ \hline \end{array}$$

10 × 100 = 1,000
10 × 20 = 200
10 × 8 = 80
6 × 100 = 600
6 × 20 = 120
6 × 8 = + 48
_____
2,048 ← Add the partial products.

The product is 2,048.

Use partial products to solve.

1. 37
   × 56
   ____

2. 416
   × 38
   ____

Use partial products to solve.

3. 472
   × 83
   ———

4. 508
   × 27
   ———

5. A mural is painted on a rectangular wall that is 154 feet long and 16 feet tall. What is the area of the wall?

   _____ square feet

6. A basketball costs $78. A team orders 25 basketballs. How much does the team spend for the basketballs?

   _____

7. A rectangular cornfield measures 358 meters long and 64 meters wide. What is the area of the cornfield?

   _____ square meters

8. A T-shirt company ships out 250 boxes of shirts each week. Each box holds 48 shirts. How many shirts does the company ship out each week?

   _____ shirts

With your child, measure the length and width of a rectangular room or wall in your home in inches. Then have them find the area of the room or wall using partial products. Have your child explain how he or she calculated the area.

Student Practice Book

Lesson **5-6**
Additional Practice

Name _____

> **Review**
>
> **You can use an algorithm to multiply a multi-digit factor and a single-digit factor.**
>
> Find the product 2,234 × 6. Use the standard algorithm for multiplication.
>
> $$\begin{array}{r} {\scriptstyle +1+2+2} \\ 2{,}234 \\ \times 6 \\ \hline 13{,}404 \end{array}$$
>
> The product is 13,404.

Solve using the standard algorithm for multiplication.

1. 478
   × 7

2. 791
   × 9

3. 384
   × 5

4. 1,477
   × 4

5. 3,519
   × 6

6. 7,568
   × 8

Solve using the standard algorithm for multiplication.

7. 296
    × 2
    ———

8. 517
    × 7
    ———

9. 3,107
    ×   9
    ———

10. 416
    × 4
    ———

11. 5,277
    ×   8
    ———

12. 2,673
    ×   5
    ———

Estimate the product. Then solve.

13. Mona earns $294 each week at her part-time job. How much does she earn working for 4 weeks?

14. Along a certain bus route, an average of 1,382 people ride the bus each day. How many riders is this in one week?

15. The average attendance at a town's high school football games is 3,856 people per game. How many people attended the team's 6 home games this season?

Make a set of 10 number cards, each with one of the digits 0 through 9. Choose 4 or 5 of the digits to form either a 3- or 4-digit number and multiply it by the remaining single-digit number using the algorithm. Try to arrange the numbers to make the greatest possible product and the least possible product. Then replace the number cards and play again.

Student Practice Book

Lesson 5-7
Additional Practice

Name _____

Review

You can use the standard algorithm to multiply 3-digit and 4-digit factors by a 2-digit factor.

Find 2,186 × 42 using the standard algorithm for multiplication.

```
        2, 1 8 6
      ×      4 2
      ─────────
        4, 3 7 2     Multiply 2,186 by 2.
     + 87, 4 4 0     Multiply 2,186 by 40.
      ─────────
       91, 8 1 2     Add the partial products.
```

The product is 91,812.

Find the product using the standard algorithm.

1. 251
   × 27

2. 974
   × 34

3. 4,109
   × 19

4. 865
   × 24

5. 376
   × 83

6. 3,489
   × 51

Student Practice Book

Solve each problem.

7. Wanda makes bracelets to sell at craft shows. Each bracelet she makes uses 38 beads. This year, Wanda has sold 351 bracelets. How many beads did she use to make the bracelets?

   _____ beads

8. Lorenzo saves $136 each week. How much money will he have saved after 1 year (52 weeks)?

   _____

9. A furniture company provides 84 screws for customers to use to put one bed together. If the company sells 511 beds, how many screws did they provide?

   _____ screws

10. Carla averages 8,275 steps each day. At this rate, how many steps will Carla walk in two weeks?

    _____ steps

Make a set of 10 number cards, each with one of the digits 0 through 9. Choose 5 or 6 of the digits to form either a 3- or 4-digit number and multiply it by a number formed by the remaining two numbers. Try to arrange the numbers to make the greatest possible product and the least possible product. Then replace the number cards and play again.

Lesson **6-1**

Additional Practice

Name _____

> **Review**
>
> **You can multiply a decimal by a power of 10.**
>
> There are 6.3×10^3 people at the football game. How many people are at the game?
>
> The exponent 3 tells you by how many factors of 10 to multiply the number. Multiply 6.3 by three factors of 10.
>
> $6.3 \times 10^3 = 6.3 \times 10 \times 10 \times 10 = 6,300$
>
> There are 6,300 people at the football game.

Write the multiplication using factors of 10. Then find the value.

1. 1.8×10^3

2. 6.4×10^2

3. 3.7×10^4

4. 5.9×10^3

5. The distance between two cities is about 1.4×10^3 miles. About how many miles apart are the two cities?

Use patterns to help you find the value of each expression.

6. $5.2 \times 10^2 =$ _____
   $5.2 \times 10^3 =$ _____
   $5.2 \times 10^4 =$ _____

7. $9.7 \times 10^1 =$ _____
   $9.7 \times 10^2 =$ _____
   $9.7 \times 10^3 =$ _____

8. $2.6 \times 10^3 =$ _____
   $2.6 \times 10^4 =$ _____
   $2.6 \times 10^5 =$ _____

9. $6.1 \times 10^2 =$ _____
   $6.1 \times 10^3 =$ _____
   $6.1 \times 10^4 =$ _____

10. The diameter of the Earth at the equator is about 7.9×10^3 miles. About how many miles is the diameter of the Earth?

11. Rosa hiked 1.3×10^3 meters before stopping for a water break. Alvin hiked 9.4×10^2 meters before stopping for water. Who hiked farther before stopping? How do you know?

12. Kenji is running in a 10K race. The course covers a total distance of 1×10^4 meters. After one hour, Kenji has run 3.2×10^3 meters. How much farther does Kenji have to run to complete the race? Write the answer in standard form and as a decimal multiplied by a power of 10.

Use an atlas or other source to find distances between two cities. Have your child write the distances in standard form and as a decimal multiplied by a power of 10. Have your child explain how the two numbers are equal.

Student Practice Book

Lesson 6-2

Additional Practice

Name _____

> ### Review
>
> **You can use estimation to determine whether a solution is reasonable.**
>
> One pound of almonds costs $4.79. Ebony buys 5.3 pounds of almonds. The cashier charges her $253.87. Should Ebony question the amount or just pay it?
>
> Ebony can estimate the cost of the almonds.
>
> 4.79 rounds to 5. 5.3 rounds to 5. Since 5 × 5 = 25, Ebony should pay about $25 for the almonds.
>
> Ebony can also find a range of reasonable costs. 4.79 is between 4 and 5. 5.3 is between 5 and 6. So a reasonable range is between 4 × 5 and 5 × 6. She will spend between $20 and $30 for the almonds.
>
> Since the amount charged, $253.87, is not in the reasonable range, Ebony should question the amount.

Estimate each product by rounding. Show your work.

1. 4.18 × 6.86

2. 2.73 × 5.17

3. 3.6 × 9.8

4. 4.55 × 7.2

Student Practice Book

Estimate each product by finding a range. Show your work.

5. 2.63 × 7.2

6. 5.56 × 1.88

7. 6.7 × 9.8

8. 4.1 × 4.58

9. Find a range of reasonable estimates for the product 7.34 × 4.78. Explain how you found the range.

10. Gasoline costs $2.86 at the filling station. Barb fills her car's tank with 8.73 gallons of fuel. About how much should Barb expect to pay for the gas? Explain which estimation strategy you used.

11. Evelyn has $30 to spend on lunch meat for a family picnic. The lunch meat costs $5.79 per pound. She estimates that she will need 7.25 pounds. Does she have enough money to buy all the lunch meat she needs? Explain how you know.

Math @ Home Activity

Looks at the prices of different items your family typically buys at the grocery store. You can use a weekly flyer or an Internet advertisement. Have your child estimate the price of buying more than 1 of the different items. Ask him or her to explain how the estimation was determined.

Student Practice Book

Lesson 6-3
Additional Practice

Name _____

Review

You can use decimal grids to solve multiplication equations involving whole numbers and decimals or decimals and decimals.

Hue has 0.7 pound of peanuts. He eats 0.3 of the peanuts for a snack. What is the weight of the peanuts that Hue eats?

Write an equation: $0.3 \times 0.7 = w$

Use a decimal grid to solve.

Shade 7 tenths of the grid.   Shade 3 tenths of the 7 tenths.

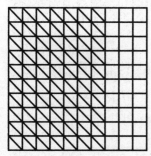 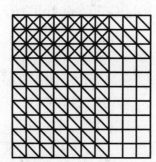

There are 21 hundredths of the whole shaded.

So Hue eats 0.21 pound of the peanuts.

Write an equation and use a decimal grid to help you solve.

1. Abbey uses 0.14 gallon of water to fill a container. She fills the container 6 times throughout the day. How many gallons of water does Abbey use in all?

Student Practice Book
55

2. In a group of students, 0.3 of the students are wearing a blue shirt. Of those students, 0.8 of the shirts have a stripe pattern. What part of the group of students are wearing a blue-striped shirt?

Complete each equation.

3. 6 × 0.3 = _____

4. 0.7 × 0.5 = _____

5. 0.16 × 6 = _____

6. 0.1 × 0.9 = _____

7. Eight concrete blocks are stacked to build a wall. If each concrete block is 0.7 foot tall, how tall is the wall?

8. Antonio is buying some apples. Each apple weighs 0.26 pound. If Antonio buys 7 apples, what is the weight of the apples?

9. Jade buys 6 packets of seeds. Each seed packet costs $0.30. She uses the representation below to find the total cost. Is her representation correct? Explain. What is Jade's total cost?

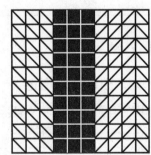 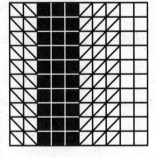

Find several common household objects, such as scissors or a juice glass. Use a scale to weigh (if available—otherwise simply estimate) each object, in pounds, and then have your child find the weight of different numbers of these objects. For example, estimate the weight of a scissors to be 0.16 pound. Then 4 pairs of scissors weighs 4 × 0.16 = 0.64 pound.

Student Practice Book

Lesson 6-4

Additional Practice

Name

Review

You can use an area model to find the product of two decimal factors or a whole number and a decimal.

Shay has a stack of 12 books. Each book weighs 2.8 pounds. How much does the stack of books weigh?

Write an equation to represent the problem: $12 \times 2.8 = w$.

Use an area model and the partial products to solve the problem.

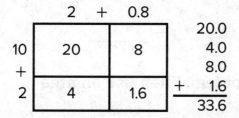

```
       20.0
        4.0
        8.0
    +   1.6
       33.6
```

The stack of books weighs 33.6 pounds.

Write an equation to represent the problem. Then use an area model to solve.

1. A farmer buys 13 rubber mats to place on the floor of his barn. Each mat is 4.5 inches thick. What is the total thickness of the mats?

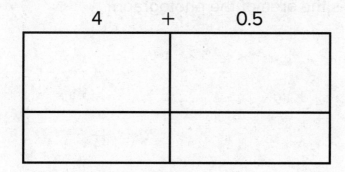

Student Practice Book

2. Aldo plants flowers in a rectangular flower bed. The flower bed is 4.7 meters long and 2.4 meters wide. What is the area of the flower bed?

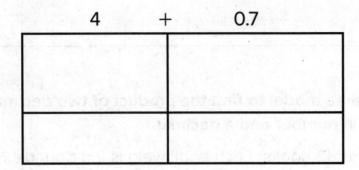

Use an area model to solve.

3. 24 × 2.3 = _____

4. 6.7 × 3.2 = _____

5. 5.6 × 19 = _____

6. 8.4 × 7.3 = _____

Write an equation to represent each problem. Then use an area model to solve.

7. Franco's car can travel about 18.3 miles per gallon. How many miles can Franco drive if his car has 14 gallons of gas in the tank?

8. A rectangular photograph measures 4.5 inches wide and 6.4 inches long. What is the area of the photograph?

Draw an area model with your child. Look for situations around your home where it would be natural to multiply decimal factors, such as finding the total weight of 12 boxes of food that each weigh 3.5 ounces. Have your child use the area model to find the total weight of the boxes. Repeat the activity with different items.

Lesson 6-5

Additional Practice

Name

Review

You can use place value to help make generalizations about multiplying decimals.

Solve each of the equations.

$35 \times 19 = m$ $35 \times 1.9 = n$ $35 \times 0.19 = p$

You know $35 \times 19 = 665$. You can use place value to help you find the solutions to the other equations.

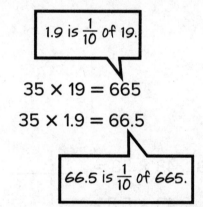
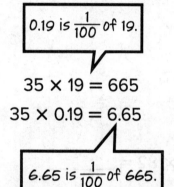

Complete each sentence.

1. 7.4 is _____ of 74. So, 7.4 × 55 is _____ of the product 74 × 55.

2. 0.26 is _____ of 26. So, 0.26 × 68 is _____ of the product 26 × 68.

3. 5.2 is _____ of 52 and 4.7 is _____ of 47. So, 5.2 × 4.7 is

   _____ of the product 52 × 47.

What is the product?

4. 65 × 23 = 1,495
   65 × 2.3 = _____
   65 × 0.23 = _____

5. 27 × 29 = _____
   27 × 2.9 = 78.3
   2.7 × 2.9 = _____

6. 84 × 37 = 3,108
   84 × 3.7 = _____
   84 × 0.37 = _____

What is the product? Explain how you can use place-value patterns to solve.

7. 36 × 82 = _____
   36 × 8.2 = _____
   3.6 × 8.2 = _____

8. 43 × 58 = _____
   43 × 5.8 = _____
   43 × 0.58 = _____

9. A class makes a mural on a wall that is 35 feet long and 12 feet tall. A copy of the mural is made on a smaller surface 3.5 feet long and 1.2 feet tall. Find the area of each mural. How does the area of the smaller mural compare to the area of the larger mural?

Have your child generate two random numbers from 1 to 99 and use the two numbers to write a multiplication equation with the answer. Then have him or her write the related equations such that the first factor is 1/10 of the first original factor, a second equation in which the second factor is 1/10 of the second original factor, and a third equation in which both factors are 1/10 of the original factors. Repeat as time **allows**.

Student Practice Book

Lesson **6-6**
Additional Practice

Name _____

Review

You can use different methods to multiply decimals.

Use a decimal grid to solve $0.6 \times 0.3 = p$.	Use partial products to solve $0.7 \times 3.4 = n$.
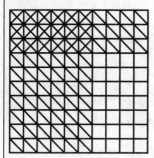 So $0.6 \times 0.3 = 0.18$.	Partial Products: $0.7 \times 3 = 2.1$ and $0.7 \times 0.4 = 0.28$ Total Product: $2.1 + 0.28 = 2.38$ So $0.7 \times 3.4 = 2.38$

Use an area model and partial products to solve 4.5×6.3.

	6	0.3
4	$4 \times 6 = 24$	$4 \times 0.3 = 1.2$
0.5	$0.5 \times 6 = 3$	$0.5 \times 0.3 = 0.15$

```
   24
   1.2
   3
+  0.15
------
  28.35
```

So $4.5 \times 6.3 = 28.35$.

What is the product?

1. 0.9×0.4

2. 6.7×2.3

Student Practice Book
61

Solve each problem. Explain the strategy used to solve.

3. Ursula walked 0.8 mile yesterday. She walked three times as far today. How far did Ursula walk today?

4. Bill jogged 4.8 miles on Saturday. On Monday, he jogged 0.6 of that distance. How many miles did Bill jog on Monday?

5. A rectangular picture frame measures 2.6 feet long and 1.6 feet tall. What is the area that is enclosed by the frame?

6. Each orange in a bag of oranges weighs about 0.3 pounds. Giselle buys a bag that contains 14 oranges. About how much does the bag of oranges weigh?

7. A rectangular vegetable garden measures 13 meters long and 9.4 meters wide. What is the area of the vegetable garden?

Have your child weigh or measure different objects around the home. Choose different numbers of objects and have your child find either the total weight of that many objects or the total length of that many objects placed end-to-end.

Student Practice Book

Lesson **7-1**
Additional Practice

Name _____

> **Review**
>
> **You can use basic facts and patterns with factors of 10 to help you divide.**
>
> Start with the basic fact 15 ÷ 3 = 5.
>
> | Multiply the dividend and the divisor by the same number of factors of 10. The quotient remains the same. | Multiply the dividend by a number of factors of 10. Keep the divisor the same. Then the quotient is multiplied by the same number of factors of 10. |
> |---|---|
> | 150 ÷ 30 = 5 | 150 ÷ 3 = 50 |
> | 1,500 ÷ 300 = 5 | 1,500 ÷ 3 = 500 |
> | 15,000 ÷ 3,000 = 5 | 15,000 ÷ 3 = 5,000 |

Write the basic fact for the division. Then use a pattern to find the quotient.

1. 3,000 ÷ 60

2. 15,000 ÷ 50

3. 180 ÷ 90

4. 21,000 ÷ 700

Complete the pattern.

5. 27 ÷ _____ = 9

    270 ÷ 30 = _____

    _____ ÷ 30 = 90

    27,000 ÷ _____ = _____

6. 36 ÷ 9 = _____

    _____ ÷ 90 = _____

    3,600 ÷ 90 = _____

    36,000 ÷ _____ = _____

Find the quotient.

7. 640 ÷ 80

8. 45,000 ÷ 500

9. 3,500 ÷ 7

10. 28,000 ÷ 400

11. There are 32,000 quarters in rolls of 40. How many rolls of quarters are there?

    _____ rolls of quarters

12. A hotel has 24,000 square feet of space on each floor to use for 60 identical rooms on each floor. What is the area of each room?

    _____ square feet

Math @ Home Activity

With your child, make some division fact cards. Turn them facedown, and select one. Then use the fact to create four more division equations, using factors of 10. After writing the four equations, select a different card and continue.

Lesson **7-2**

Additional Practice

Name

> **Review**
>
> **You can use compatible numbers to estimate quotients.**
>
> A scout troop collected 2,854 cans over the last 36 days. About how many cans were collected each day?
>
> To solve, estimate the quotient 2,854 ÷ 36.
>
> Look for compatible numbers that are easy to divide mentally.
>
> 2,854 is close to 2,800.   2,854 is close to 2,700.
>
> 36 is close to 40.   36 is close to 30.
>
> A possible estimate is   A possible estimate is
> 2,800 ÷ 40 = 70.   2,700 ÷ 30 = 90.
>
> The scout troop collected   The scout troop collected about 90
> about 70 cans each day.   cans each day.

Estimate the quotient. Show how you made the estimate.

1. 3,000 ÷ 48

2. 4,000 ÷ 74

3. 360 ÷ 35

4. 2,100 ÷ 63

Student Practice Book
65

Estimate the quotient. Show how you made the estimate.

5. 4,214 ÷ 78

6. 1,744 ÷ 63

7. 8,070 ÷ 84

8. 3,316 ÷ 42

9. 24,385 ÷ 47

10. 22,196 ÷ 74

11. A football stadium holds 63,588 people. The stadium has 84 sections. About how many people sit in each section?

about _____ people

12. A salesman figured out that he has traveled 4,755 miles over the last 8 months. About how many miles did the salesman travel each month?

about _____ miles

With your child, make a set of number cards with the digits 0 to 9. Turn the cards facedown and choose six of the cards. Arrange them into a division problem of a 4-digit number divided by a 2-digit number, and estimate the quotient. Then rearrange the numbers to try and obtain a division problem that has a greater estimate, and then a lesser estimate. Then return the cards to the set, mix them, and repeat the activity.

Student Practice Book

Lesson **7-3**
Additional Practice

Name _____

Review

You can use multiplication to help you find a quotient.

On a bike trip, the riders rode 345 miles over 15 days. How many miles did they ride each day?

To solve, find the quotient $345 \div 15 = d$.

Write a related multiplication equation: $d \times 15 = 345$. Find how many groups of 15 there are in 345.

$10 \times 15 = 150$	345 − 150 195
$10 \times 15 = 150$	195 − 150 45
$3 \times 15 = 45$	45 − 45 0

There are $10 + 10 + 3 = 23$ groups of 15. So $345 \div 15 = 23$.

The riders rode 23 miles each day.

1. How many groups of 18 can you make from 270?

2. How many groups of 22 can you make from 462?

3. How many groups of 13 can you make from 364?

4. How many groups of 34 can you make from 544?

Write the related multiplication equation. Then solve.

5. $442 \div 17 = n$

6. $473 \div 11 = w$

7. $456 \div 24 = m$

8. $325 \div 13 = b$

9. A landscaper plants 288 flowers. The flowers are planted in 18 equal rows. How many flowers are in each row?

   _____ flowers

10. A farmer has 209 chickens. He builds enough coops so that there can be 11 chickens in each coop. How many coops does the farmer build?

    _____ coops

With your child, make two sets of number cards, one set with the numbers from 5 to 19, and the second set with the numbers 11 to 40. Have your child select one card from each set. Away from your child, multiply the two numbers, and give your child one of the numbers and the product. Have him or her find the number of groups on the first card that are in the product. The answer will be the number on the second card. Have your child explain how he or she found the answer. After verifying that the answer is correct, return the cards and repeat the activity by selecting again.

Student Practice Book

Lesson 7-4

Additional Practice

Name _____

> **Review**
>
> **You can use multiplication to help you find a quotient. An area model can help you to keep track of the partial products.**
>
> The area of a rectangle is 1,564 square feet. The width of the rectangle is 34 feet. What is the length?
>
> To solve, find the quotient $1{,}564 \div 34 = l$. Use an area model.
>
> 34
>
> | | |
> |---|---|
> | 30 × 34 = 1,020 | 30 |
> | 10 × 34 = 340 | 10 |
> | 5 × 34 = 170 | 5 |
> | 1 × 34 = 34 | + 1 |
> | | 46 |
>
> The partial products add to 1,564. The factors used add to equal the quotient of 46.
>
> The length of the rectangle is 46 feet.

Find the quotient. Use an area model to solve.

1. $966 \div 42$

2. $1{,}764 \div 28$

Find each quotient. Use an area model to solve.

3. 1,760 ÷ 32 = _____

4. 748 ÷ 22 = _____

5. 2,961 ÷ 63 = _____

6. 4,482 ÷ 54 = _____

7. The area of a rectangular wall is 882 square feet. The wall is 18 feet tall. How long is the wall?

   _____ feet

8. The playground at a school is in the shape of a rectangle. The playground covers an area of 6,364 square feet. The playground is 86 feet long. What is the width of the playground?

   _____ feet

With your child, make a set of number cards using the digits 1 through 9. Have your child select two of the cards to form a 2-digit number. Then you select two more cards and form another 2-digit number. Away from your child, multiply the two 2-digit numbers, and tell your child the product. Have your child find the number that you formed from the two cards you selected. Have your child explain how they found the answer. Return the cards to the set and repeat the activity.

Student Practice Book

Lesson 7-5
Additional Practice

Name _____

Review

You can use the partial quotients algorithm to help you find a quotient.

The area of a rectangular field is 3,216 square feet. The width of the field is 48 feet. What is the length?

To solve, find the quotient $3,216 \div 48 = l$. Use the partial quotients algorithm.

The factors used, shown along the side, add to equal the quotient, 67.

The length of the field is 67 feet.

```
    48)3,216
       -2,400  | 50
         816
       -  480  | 10
         336
       -  240  | 5
          96
       -   96  | 2
           0
```

Find the quotient. Use the partial quotients algorithm to solve.

1. $832 \div 26 = $ _____

2. $3,648 \div 48 = $ _____

Student Practice Book

Find the quotient. Use the partial quotients algorithm to solve.

3. 518 ÷ 14 = _____

4. 756 ÷ 36 = _____

5. 3,285 ÷ 45 = _____

6. 6,512 ÷ 74 = _____

7. Madeleine rides her bicycle around a track. She rides 14 laps and counts 1,372 times she pushes on each pedal. How many times does Madeleine push on each pedal for one lap?

_____ pushes

With your child, make a set of number cards using the digits 1 through 9. Have your child select two of the cards to form a 2-digit number. Then you select two more cards and form another 2-digit number. Away from your child, multiply the two 2-digit numbers, and tell your child the product. Have your child find the number that you formed from the two cards you selected. Have your child use partial quotients to explain how they found the answer. Return the cards to the set and repeat the activity.

Student Practice Book

Lesson **7-6**

Additional Practice

Name _____

Review

You can use the partial quotients algorithm to help you find a quotient and any remainder.

Jerry has 275 marbles. He places them into bags with 16 marbles in each bag. How many bags will Jerry have? How many marbles will he have left over?

To solve, find the quotient $275 \div 16 = n$. Use the partial quotients algorithm.

```
  16 ) 275
     - 160  | 10
       115
     - 112  |  7
         3  |
```

The factors used, shown along the side, add to be the quotient, 17. The 3 at the bottom is the remainder.

Jerry will have 17 bags of marbles with 3 marbles left over.

Use the partial quotients algorithm to solve.

1. $607 \div 17 = $ _____

2. $3{,}766 \div 52 = $ _____

Student Practice Book

Use the partial quotients algorithm to solve.

3. 834 ÷ 23 = _____ **4.** 2,512 ÷ 49 = _____

5. 2,267 ÷ 82 = _____ **6.** 4,441 ÷ 65 = _____

7. One day at the apple orchard, 2,788 apples were picked. They were placed into bags with 32 apples in each bag. How many bags were made? How many apples were left over?

_____ bags made; _____ apples left over

With your child, make a set of number cards using the digits 0 through 9. Have your child select two of the cards to form a 2-digit number and then four more cards to form a 4-digit number. Have your child find the quotient and remainder when the 4-digit number is divided by the 2-digit number using the partial quotients algorithm. Then replace the cards and choose new numbers to repeat the activity.

Student Practice Book

Lesson 7-7
Additional Practice

Name _____

Review

You can use the partial quotients algorithm to help you find a quotient and any remainder. Then you can interpret the remainder in the context of the problem to answer the question.

There are 325 chairs to be set up in the gym for a performance. There are to be 18 chairs in each row. How many rows will be needed to arrange all of the chairs?

To solve, find the quotient $325 \div 18 = n$. Use the partial quotients algorithm.

```
   18 ) 335
      - 180 | 10
        155
      - 144 |  8
         11
```

The factors used, shown along the side, add to be the quotient, 18. The 11 at the bottom is the remainder.

There will be 18 rows of 18 chairs and 1 row of 11 chairs.

So 19 rows of chairs are needed to arrange all of the chairs.

1. Roger reviews movies for the local newspaper. He is given a budget of $350 and can spend $16 at each theater. How many movies can he review? Explain your answer.

Student Practice Book

2. At the farm one morning, 295 eggs are collected. They are packaged into containers of 12 (one dozen). How many dozen cartons are filled? Explain your answer.

3. A sporting goods company produces 3,400 tennis balls each day. They ship out cartons with 36 tennis balls in each carton. After packing and shipping out as many full cartons as possible, how many tennis balls will be left to ship out the next day? Explain your answer.

4. A farmer wants to plant 2,750 corn stalks in rows of 84 corn stalks in each row. How many rows will the farmer need in order to plant all of the corn stalks? Explain your answer.

With your child, find some items around your home that are numerous and can be divided into smaller groups, such as a container of screws or nails, or a bag or jar of nuts. Estimate the number of total items, and think of a scenario in which the items have to be divided into equal groups. Ask questions so that the remainder has to be interpreted in different ways.

Student Practice Book

Lesson **8-1**
Additional Practice

Name _____

> **Review**
>
> **You can use the relationship between place-value positions to divide by decimals and powers of 10.**
>
> | Divide by decimals | Divide by powers of 10 |
> |---|---|
> | 48.5 ÷ 1 = 48.5 | 48.5 ÷ 1 = 48.5 |
> | 48.5 ÷ 0.1 = 485 | 48.5 ÷ 10 = 4.85 |
> | 48.5 ÷ 0.01 = 4,850 | 48.5 ÷ 100 = 0.485 |
> | When dividing by decimals, the quotient has to be greater than the dividend. So shift the digits to the left of the decimal point to make the number greater. | When dividing by powers of 10, the quotient has to be less than the dividend. So shift the digits to the right of the decimal point to make the number less. |

Complete the pattern of quotients.

1. 29.7 ÷ 100 = _____
   29.7 ÷ 10 = _____
   29.7 ÷ 1 = _____
   29.7 ÷ 0.1 = _____
   29.7 ÷ 0.01 = _____

2. 8.3 ÷ 100 = _____
   8.3 ÷ 10 = _____
   8.3 ÷ 1 = _____
   8.3 ÷ 0.1 = _____
   8.3 ÷ 0.01 = _____

Student Practice Book

Find the quotient.

3. $41.6 \div 0.1 =$ _____

4. $7.8 \div 0.01 =$ _____

5. $30.4 \div 10 =$ _____

6. $38.2 \div 100 =$ _____

7. $4.2 \div 100 =$ _____

8. $207.8 \div 10 =$ _____

9. $26.4 \div 0.01 =$ _____

10. $4.8 \div 0.1 =$ _____

11. Arabella has $13, all in dimes. How many dimes does she have? Explain.

12. Jorge has $22.50, all in pennies. How many pennies does he have? Explain.

13. Henry walks to and from school each day. After 100 days of school, he has walked 125 miles. How many miles does Henry walk to and from school each day? Explain.

Find some prices of items around your home, in a store, or in an ad. Have your child tell how many dimes it would take to make that amount. Then have your child tell how many pennies it would take to make that amount.

Student Practice Book

Lesson 8-2
Additional Practice

Name _____

Review

You can use compatible numbers to estimate a quotient.

Estimate the quotient 31.6 ÷ 0.6. When dividing by a decimal less than 1, multiply the dividend and divisor so that the divisor is a whole number, then look for compatible numbers.	Estimate the quotient 31.6 ÷ 8.4. When dividing by a decimal greater than 1, look for compatible numbers.
To make 0.6 into a whole number, multiply by 10. Also, multiply the dividend, 31.6, by 10. 31.6 ÷ 0.6 316 ÷ 6	An estimate for 31.6 ÷ 8.4 is 32 ÷ 8 = 4.
Now use compatible numbers. An estimate for 316 ÷ 6 is 300 ÷ 6 = 50. So 31.6 ÷ 0.6 is about 50.	So 31.6 ÷ 8.4 is about 4.

Which is the quotient?

1. 31.28 ÷ 0.46
   a. 0.68
   b. 6.8
   c. 68
   d. 680

2. 43.99 ÷ 8.3
   a. 0.53
   b. 5.3
   c. 53
   d. 530

Estimate the quotient. Show your work.

3. 24.45 ÷ 0.8

4. 73.4 ÷ 7.7

5. 12.83 ÷ 0.21

6. 9.1 ÷ 2.8

7. 83.24 ÷ 9.06

8. 65.2 ÷ 0.87

9. Harriet spends $12.58 on some stickers. Each sticker costs $0.06. About how many stickers did Harriet buy? Explain.

10. A bicycle race covers a distance of 64.5 kilometers. There are water stations every 7.6 kilometers. About how many water stations are there along the course? Explain.

Practice estimating quotients with your child. While preparing dinner or packing lunches, look for situations where it would be natural to find an estimate for a quotient. For example, if a jar of peanut butter contains 11.7 ounces, and each sandwich uses 0.38 ounce, about how many sandwiches can be made? Allow them to use a calculator to check the estimate

Student Practice Book

Lesson 8-3
Additional Practice

Name Sheyla Rodriguez Casillas

Review

You can use tenths or hundredths grids to show how to divide a decimal by a whole number.

Molly has $2.60. She wants to divide the money into 4 equal groups. How much money will be in each group?

Use hundredths decimal grids to represent the division.

Divide 260 hundredths into 4 equal groups. Each group has 65 hundredths.

There will be $0.65 in each group.

Write the equation shown by the decimal grids.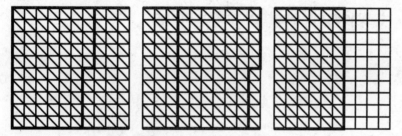

1.

2.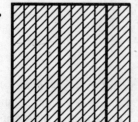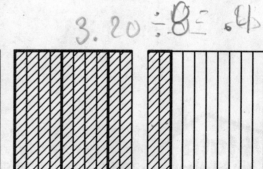

Student Practice Book
81

What is the quotient? Use decimal grids to solve.

3. 4.2 ÷ 6 = .7

4. 1.86 ÷ 3 = .62

5. 3.36 ÷ 8 = .42

6. 4.32 ÷ 4 = 1.08

7. 1.8 ÷ 9 = .2

8. 6.37 ÷ 7 = .91

9. Miriam has a length of string that is 6.5 inches long. She cuts it into 5 equal lengths. How long is each piece of string?

    __1.3__ inches

10. Along a hike, Jon takes 6 pictures. He takes 1 picture at equal distances along the 3.72-mile trail. How far did Jon hike between pictures?

    __.62__ mile(s)

On a sheet of paper, write a similar problem to the ones in this lesson involving the division of a decimal by a whole number. Have your child show you how to use decimal grids to find the quotient. Repeat with a different division problem.

Student Practice Book

Lesson 8-4
Additional Practice

Name Sheyla Rodriguez Casillas

Review

You can use place-value understanding and equivalent representations to divide a decimal by a whole number.

Michael has a board that is 1.44 meters long. He wants to cut it into 3 pieces of equal length. How long should Michael cut each board?

Write a division equation to represent the problem.

$$1.44 \div 3 = b$$

Write an equivalent representation.

144 hundredths ÷ 3 = b

144 hundredths ÷ 3 = 48 hundredths

Each board will be 0.48 meters long.

---

1. Which is an equivalent representation of 3.5 ÷ 5?

   A. 35 tens ÷ 5
   B. 35 ones ÷ 5
   (C.) 35 tenths ÷ 5
   D. 35 hundredths ÷ 5

2. Which is an equivalent representation of 2.16 ÷ 4?

   A. 216 tens ÷ 4
   B. 216 ones ÷ 4
   C. 216 tenths ÷ 4
   (D.) 216 hundredths ÷ 4

Write an equivalent representation for the division. Then find the quotient.

3. 1.86 ÷ 6 = .62

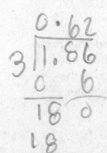

4. 0.72 ÷ 4 .18

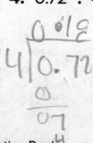

Student Practice Book
83

Write an equivalent representation for the division. Then find the quotient.

5. $5.6 \div 8$

6. $4.32 \div 3$

7. $1.15 \div 5$

8. $14.8 \div 4$

9. Greta buys 5 pens for $3.45. How much does each pen cost? Explain how you can use an equivalent representation to help you solve.

10. Jack buys 8 pounds of apples for $9.52. How much does 1 pound of apples cost?

11. A length of ribbon is 0.8 meter long. Justine cuts the ribbon into 4 equal lengths to wrap presents. How long is each piece of ribbon?
    _____ meter

With your child, be alert to examples of decimal numbers around your home, at the store, or just in your everyday experiences. Suggest a whole-number divisor, and have your child tell you an equivalent representation of the division, and tell (or estimate) the quotient. Repeat for other examples.

Lesson 8-6

Additional Practice

Name _Sheyla Rodriguez Casillas_

Review

You can use powers of 10 to help you divide a decimal number by another decimal number.

The area of a rectangular garden is 51.2 square meters. The width of the garden is 6.4 meters. What is the length of the garden?

Write a division equation to represent the problem.

$$51.2 \div 6.4 = l$$

Use a power of 10 so that the divisor, 6.4, is a whole number:

$6.4 \times 10 = 64$

Multiply the dividend by the same power of 10: $51.2 \times 10 = 512$

Write an equivalent equation with the new dividend and divisor:

$512 \div 64 = h$

Since $512 \div 64 = 8$, it must be that $51.2 \div 6.4 = 8$.

The garden is 8 meters long.

Write an equivalent division so that the divisor is a whole number. Then find the quotient.

1. $2.7 \div 0.9$

2. $3.2 \div 0.4$

3. $9.4 \div 4.7$

4. $24.94 \div 0.58$

Student Practice Book

Write an equivalent division so that the divisor is a whole number. Then find the quotient.

5. 16.8 ÷ 4.2

6. 14.35 ÷ 0.35

7. 33.06 ÷ 0.87

8. 170.1 ÷ 2.7

9. Dexter is designing a rectangular bedroom. The area of the bedroom is to be 137.5 square feet, and the length is to be 12.5 feet. What is the width of the bedroom?

    _____ feet

10. Amelia has 0.96 pound of grated cheese. She uses 0.06 pound of cheese on each of her salads. How many salads can Amelia make?

    _____ salads

11. A fence post is placed every 4.2 feet. How many fence posts are needed for a fence that is 176.4 feet long?

    _____ fence posts

Look for situations around your home where dividing two decimals would be natural. Ask your child to identify a situation and explain how to find the quotient. For example, if there are 2.5 pounds of strawberries in a container and they are divided in 0.5-pound bags, how many bags can be filled?

Student Practice Book

Lesson **9-1**

Additional Practice

Name _Sheyla Rodriguez Casillas._

> **Review**
>
> You can use the benchmarks 0, $\frac{1}{2}$, and 1 to estimate sums and differences of fractional amounts.
>
> Archie uses $\frac{1}{4}$ foot of blue-striped ribbon and $\frac{4}{5}$ foot of solid blue ribbon. He says that he uses $\frac{5}{9}$ foot of ribbon in all. Is Archie's estimate reasonable?
>
>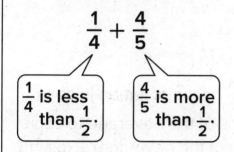
>
> The sum should be close to 1. Archie's sum of $\frac{5}{9}$ is a little more than $\frac{1}{2}$, so his estimate is not reasonable.

1. Match the sum or difference with its best estimate.

   $\frac{1}{3} + \frac{7}{8}$ close to 0

   $\frac{9}{10} + \frac{5}{6}$ close to $\frac{1}{2}$

   $\frac{8}{9} - \frac{2}{3}$ close to 1

   $\frac{6}{7} - \frac{1}{4}$ close to 2

Student Practice Book
89

Use estimation with benchmark fractions to answer each question.

2. Kelli swims $\frac{2}{3}$ mile in the morning and $\frac{1}{6}$ mile in the evening. About how many mile(s) does Kelli swim in one day? Explain.

3. Garry practices playing piano for $\frac{3}{5}$ hour on Wednesday and $\frac{1}{2}$ hour on Thursday. He says that he practices for $\frac{1}{2}$ hour more on Wednesday than on Thursday. Is he correct? Explain.

4. Norma mixes $\frac{11}{12}$ gallon of water and $\frac{4}{5}$ gallon of lemon juice to make a beverage. She needs 2 gallons of the beverage to serve at a picnic. Does Norma have enough of the beverage? Explain.

5. Wilbert's cooler holds $\frac{5}{6}$ pound of ice. He fills the cooler with $\frac{1}{5}$ pound of ice and says that the cooler is almost full. Is he correct? Explain.

Fractions are commonly used in recipes. While preparing a recipe, ask your child to estimate the values used and then find the difference between two different fractional amounts of ingredients used. Then have them estimate the sum of two fractional ingredients used. Allow your child to use different measuring cups to show his or her work.

Student Practice Book

Lesson 9-2
Additional Practice

Name Sheyla Rodriguez Casillas

Review

You can use fraction tiles to represent addition of fractions with unlike denominators.

Sandy uses $\frac{2}{3}$ gallon of milk in the morning and $\frac{1}{6}$ gallon of milk in the evening. How much of the milk does Sandy use in all?

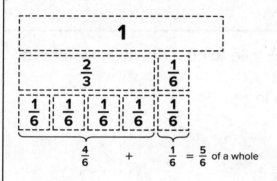

$\frac{2}{3}$ is an equivalent fraction to $\frac{4}{6}$.

Now that the fractions have like denominators, you can add them and name the sum.

Sandy uses $\frac{5}{6}$ gallon of milk.

What equation do the fraction tiles represent?

1.
   1
   | $\frac{1}{5}$ | $\frac{1}{5}$ | $\frac{1}{10}$ | $\frac{1}{10}$ | $\frac{1}{10}$ |
   | $\frac{1}{10}$ | $\frac{1}{10}$ | $\frac{1}{10}$ | $\frac{1}{10}$ | $\frac{1}{10}$ | $\frac{1}{10}$ | $\frac{1}{10}$ |

2.
   1
   | $\frac{1}{4}$ | $\frac{1}{8}$ | $\frac{1}{8}$ | $\frac{1}{8}$ |
   | $\frac{1}{8}$ | $\frac{1}{8}$ | $\frac{1}{8}$ | $\frac{1}{8}$ | $\frac{1}{8}$ |

Student Practice Book

What is the sum?

3. $\frac{1}{4} + \frac{5}{12}$	4. $\frac{1}{3} + \frac{1}{2}$
5. $\frac{1}{4} + \frac{1}{8}$	6. $\frac{7}{12} + \frac{1}{6}$
7. $\frac{2}{5} + \frac{1}{2}$	8. $\frac{7}{12} + \frac{1}{3}$

9. Kathryn runs $\frac{3}{4}$ mile. She walks $\frac{1}{8}$ mile to cool down. How far did Kathryn run and walk? _____ mile

10. Harry reads for $\frac{2}{3}$ hour. Before going to sleep, he reads for another $\frac{1}{4}$ hour. For how long did Harry read? _____ hour

11. Ned ate $\frac{5}{12}$ of the apple slices on the plate. Ted ate $\frac{1}{6}$ of the apple slices on the plate. What fraction of the apple slices did Ned and Ted eat? _____ of the apple slices

Use construction paper and scissors to create fraction tiles. Write several addition of fractions problems, similar to the ones in this lesson. Have your child use the fraction tiles to show you how to find the sum of the fractions. Look for situations around your home where it is natural to add two fractions. Link these situations to the addition problems your child solves.

Student Practice Book

Lesson 9-3
Additional Practice

Name: Sheyla Rodriguez Casillas

> **Review**
>
> You can add fractions with unlike denominators by finding a common multiple of the denominators to use as a common denominator.
>
> Ollie's math notebook weighs $\frac{3}{8}$ pound. His science notebook weighs $\frac{1}{6}$ pound. How much do Ollie's two notebooks weigh together?
>
> To solve, find $\frac{3}{8} + \frac{1}{6}$.
>
> For the denominators 8 and 6, a common multiple is 24. Write equivalent fractions using 24 as the denominator.
>
> $\frac{3}{8} \times \frac{3}{3} = \frac{9}{24}$ $\frac{1}{6} \times \frac{4}{4} = \frac{4}{24}$
>
> $\frac{3}{8} + \frac{1}{6} = \frac{9}{24} + \frac{4}{24} = \frac{13}{24}$
>
> Ollie's two notebooks weigh $\frac{13}{24}$ pound together.

Find the sum. Show your work.

1. $\frac{1}{6} + \frac{1}{3} =$ _____

2. $\frac{1}{5} + \frac{2}{3} =$ _____

3. $\frac{1}{6} + \frac{4}{9} =$ _____

4. $\frac{3}{4} + \frac{1}{6} =$ _____

5. $\frac{1}{4} + \frac{7}{10} =$ _____

6. $\frac{2}{3} + \frac{1}{4} =$ _____

Student Practice Book

Solve the problem. Show your work.

7. Elsie makes a braid from several strands of string. After one hour, the braid is $\frac{3}{5}$ foot long. During the next hour, the braid is $\frac{1}{3}$ foot longer. How long is Elsie's braid after two hours? _____ foot

8. Damon mixes $\frac{3}{4}$ gallon of white paint with $\frac{1}{6}$ gallon of blue paint to make a light blue color. How much paint does Damon have? _____ gallon

9. Josue walks $\frac{1}{8}$ mile from his house to his friend's house. Then they walk $\frac{2}{3}$ mile from his friend's house to the park. How far did Josue walk? _____ mile

10. Isabel pours $\frac{1}{12}$ gallon of drink mix into a pitcher that contains $\frac{7}{8}$ gallon of water. How much of the drink does Isabel make? _____ gallon

Practice adding fractions with unlike denominators with your child. Look for situations in which two fractions have to be added, or simply write an addition equation at the top of a sheet of paper. Have your child rewrite the equation using a common denominator, and find the sum. Then have them explain how to find the sum before moving on to another equation.

Lesson 9-4
Additional Practice

Name: Sheyla Rodriguez Casillas

Review

You can use fraction tiles to represent subtraction of fractions with unlike denominators.

Felipe has $\frac{5}{6}$ foot of string. Tristen has $\frac{1}{2}$ foot of string. How much more string does Felipe have than Tristen?

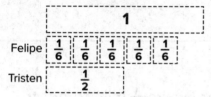

Since the denominators are different, you can create an equivalent fraction for $\frac{1}{2}$ with a denominator of 6.

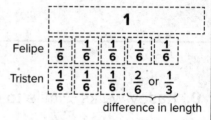

difference in length

Felipe has $\frac{2}{6}$ foot, or $\frac{1}{3}$ foot, more string than Tristen.

What subtraction equation do the fraction tiles represent?

1.

1

| $\frac{1}{5}$ | $\frac{1}{5}$ |
| $\frac{1}{10}$ | $\frac{1}{10}$ | $\frac{1}{10}$ |

2.

1

| $\frac{1}{4}$ | $\frac{1}{4}$ | $\frac{1}{4}$ |

| $\frac{1}{3}$ |

Student Practice Book
95

What is the difference?

3. $\frac{5}{12} - \frac{1}{4}$	4. $\frac{1}{2} - \frac{1}{3}$
5. $\frac{3}{4} - \frac{1}{8}$	6. $\frac{7}{12} - \frac{1}{6}$
7. $\frac{1}{2} - \frac{2}{5}$	8. $\frac{7}{12} - \frac{1}{3}$

9. Kasey walks $\frac{2}{3}$ mile to get to school. She walks $\frac{1}{4}$ mile to get to her friend's house. How much farther does Kasey walk to school than to her friend's house? _____ mile

10. Emmaline creates a tower of blocks that is $\frac{11}{12}$ foot tall. Jed's tower of blocks is $\frac{2}{3}$ foot tall. How much taller is Emmaline's tower than Jed's tower? _____ foot

Use construction paper and scissors to create fraction tiles. Write several subtraction of fractions problems, similar to the ones in this lesson. Have your child use the fraction tiles to show you how to find the difference of the fractions. Look for situations around your home where it is natural to subtract two fractions. Link these situations to the subtraction problems your child solves.

Student Practice Book

Lesson 9-5
Additional Practice

Name Sheyla Rodríguez Casillas

> **Review**
>
> **You can subtract fractions with unlike denominators by finding a common multiple of the denominators to use as a common denominator.**
>
> Ollie's math notebook weighs $\frac{3}{8}$ pound. His science notebook weighs $\frac{1}{6}$ pound. How much more does Ollie's math notebook weigh than his science notebook?
>
> To solve, find $\frac{3}{8} - \frac{1}{6}$.
>
> For the denominators 8 and 6, a common multiple is 24. Write equivalent fractions using 24 as the denominator.
>
> $\frac{3}{8} \times \frac{3}{3} = \frac{9}{24}$ $\frac{1}{6} \times \frac{4}{4} = \frac{4}{24}$
>
> $\frac{3}{8} - \frac{1}{6} = \frac{9}{24} - \frac{4}{24} = \frac{5}{24}$
>
> Ollie's math notebook weighs $\frac{5}{24}$ pound more than his science notebook.

Find the difference. Show your work.

1. $\frac{1}{3} - \frac{1}{6} = $ _____

2. $\frac{2}{3} - \frac{1}{5} = $ _____

3. $\frac{4}{9} - \frac{1}{6} = $ _____

4. $\frac{3}{4} - \frac{1}{6} = $ _____

5. $\frac{7}{10} - \frac{1}{4} = $ _____

6. $\frac{2}{3} - \frac{1}{4} = $ _____

Solve the problem. Show your work.

7. Raymundo walks $\frac{9}{10}$ mile. Mica walks $\frac{3}{5}$ mile. How much farther does Raymundo walk than Mica? _____ mile

8. Yesterday, Deborah worked for $\frac{5}{6}$ hour on her homework. Today she worked $\frac{3}{4}$ hour on her homework. How much longer did Deborah spend on her homework yesterday than today?

   _____ hour

9. A piece of string is $\frac{9}{10}$ foot long. Jody cuts the string so that one piece is $\frac{1}{2}$ foot long. How long is the other piece of string?

   _____ foot

10. The waterfall is $\frac{3}{4}$ kilometer from the nature center. Gary has hiked $\frac{2}{5}$ kilometer so far on the way to the waterfall. How much farther does Gary have to hike to get to the waterfall?

    _____ kilometer

Math @ Home Activity

Practice subtracting fractions with unlike denominators with your child. Look for situations in which two fractions have to be subtracted, or simply write a subtraction equation at the top of a sheet of paper. Have your child rewrite the equation using a common denominator, explain the steps to you, and then find the difference. Do other examples as time permits.

Student Practice Book

Lesson 9-6
Additional Practice

Name _Sheyla Rodriguez Casillas._

Review

You can add mixed numbers by adding the whole number parts and the fractional parts.

Myra walks $3\frac{1}{3}$ miles on Saturday and $4\frac{1}{4}$ miles on Sunday. How many miles does Myra walk on those two days?

To solve, find $3\frac{1}{3} + 4\frac{1}{4}$.

Decompose the addends: $3\frac{1}{3} = 3 + \frac{1}{3}$ and $4\frac{1}{4} = 4 + \frac{1}{4}$

Rewrite the sum: $3\frac{1}{3} + 4\frac{1}{4} = 3 + \frac{1}{3} + 4 + \frac{1}{4}$

Change the order of the addends so that the whole numbers are together and the fractions are together:

$3 + \frac{1}{3} + 4 + \frac{1}{4} = 3 + 4 + \frac{1}{3} + \frac{1}{4}$

Add the whole numbers: $3 + 4 = 7$

Add the fractions: $\frac{1}{3} + \frac{1}{4} = \frac{4}{12} + \frac{3}{12} = \frac{7}{12}$

Add the whole numbers and the fractions: $7 + \frac{7}{12} = 7\frac{7}{12}$

Myra walked $7\frac{7}{12}$ miles on the two days.

What is the sum? Show your work.

1. $2\frac{1}{6} + 5\frac{2}{3} = $ _____

2. $8\frac{3}{4} + 3\frac{1}{10} = $ _____

3. $6\frac{3}{5} + 4\frac{1}{3} = $ _____

4. $5\frac{1}{4} + 3\frac{2}{3} = $ _____

Student Practice Book

Solve the problem. Show your work.

5. Darcie runs $4\frac{1}{2}$ miles in the morning and $3\frac{3}{8}$ miles in the evening. How many miles does Darcie run in all?

   _____ miles

6. A frog jumps $2\frac{1}{4}$ feet. Then the frog jumps $2\frac{1}{6}$ feet. What is the total distance the frog jumped? _____ feet

7. A plant is $6\frac{1}{2}$ inches tall. After one week, the plant grows $2\frac{2}{5}$ inches. Now how tall is the plant? _____ inches

8. Clara buys two watermelons for a family picnic. One watermelon weighs $6\frac{1}{3}$ pounds. The second watermelon weighs $5\frac{5}{12}$ pounds. How much do the two watermelons weigh together?

   _____ pounds

With your child, find two objects around your home that can be placed next to each other. Measure the length of each, using a foot-long ruler, and write the measurement as a mixed number of feet. Have your child add the two measurements, and then place the two objects next to each other and measure the total length to verify the sum. Find two other objects, and repeat as time permits.

Student Practice Book

Lesson 9-7

Additional Practice

Name _____

> **Review**
>
> **You can subtract mixed numbers by subtracting the whole number parts and the fractional parts.**
>
> Sheila jogs $4\frac{1}{4}$ miles on Wednesday and $6\frac{2}{3}$ miles on Saturday. How many miles more does Sheila jog on Saturday?
>
> To solve, find $6\frac{2}{3} - 4\frac{1}{4}$.
>
> Subtract the whole number: $6\frac{2}{3} - 4 = 2\frac{2}{3}$
>
> Subtract the fraction: $2\frac{2}{3} - \frac{1}{4} = 2\frac{8}{12} - \frac{3}{12} = 2\frac{5}{12}$
>
> Sheila jogged $2\frac{5}{12}$ miles more on Saturday.

What is the difference? Show your work.

1. $5\frac{2}{3} - 1\frac{1}{2} =$ _____

2. $8\frac{3}{4} - 3\frac{1}{10} =$ _____

3. $6\frac{3}{8} - 5\frac{1}{4} =$ _____

4. $3\frac{7}{10} - 1\frac{2}{5} =$ _____

5. $10\frac{5}{6} - 7\frac{1}{4} =$ _____

6. $7\frac{7}{9} - 3\frac{1}{3} =$ _____

Student Practice Book

Solve the problem. Show your work.

7. Edison has $5\frac{5}{6}$ gallons of paint. After painting his room, he has $2\frac{1}{4}$ gallons left. How much paint does Edison use to paint his room? _____ gallons

8. Conrad has two boards. One board is $4\frac{3}{5}$ feet long. The second board is $2\frac{1}{3}$ feet long. How much longer is the first board than the second board? _____ feet

9. A hiking trail is $6\frac{5}{8}$ kilometers long. Karen has hiked $4\frac{1}{3}$ kilometers so far. How much farther does Karen have to go to reach the end of the trail? _____ kilometers

10. Paul jumps $6\frac{1}{4}$ feet. Petra jumps $8\frac{3}{5}$ feet. How much farther does Petra jump than Paul? _____ feet

Math @ Home Activity

With your child, find two objects around your home whose lengths can be compared. Measure the length of each, using a foot-long ruler, and write the measurements as a mixed number of feet. Have your child subtract the two measurements to see how much longer one object is than the other, and then place the two objects next to each other and measure the difference in lengths to verify the result. Find two other objects, and repeat as time permits.

Student Practice Book

Lesson 9-8
Additional Practice

Name _____

Review

You can add and subtract mixed numbers with regrouping.

Ken walked $2\frac{3}{4}$ miles yesterday. He walked $4\frac{2}{3}$ miles today.

How far did Ken walk both days?	How much farther did Ken walk today than yesterday?
To solve, find $2\frac{3}{4} + 4\frac{2}{3}$.	To solve, find $4\frac{2}{3} - 2\frac{3}{4}$.
Add the whole numbers: $2 + 4 = 6$	Write the fractions using a common denominator: $4\frac{8}{12} - 2\frac{9}{12}$
Add the fractions: $\frac{3}{4} + \frac{2}{3} = \frac{9}{12} + \frac{8}{12} = \frac{17}{12} = 1\frac{5}{12}$	Regroup 1 whole as $\frac{12}{12}$: $3\frac{20}{12} - 2\frac{9}{12}$
Add the whole number and the fraction: $6 + 1\frac{5}{12} = 7\frac{5}{12}$	Subtract the whole numbers and the fractions: $3\frac{20}{12} - 2\frac{9}{12} = 1\frac{11}{12}$
Ken walked $7\frac{5}{12}$ miles both days.	Ken walked $1\frac{11}{12}$ miles farther today than yesterday.

What is the sum or difference? Show your work.

1. $5\frac{1}{2} - 1\frac{2}{3} =$ _____

2. $3\frac{1}{4} + 3\frac{9}{10} =$ _____

3. $8\frac{1}{4} - 3\frac{7}{10} =$ _____

4. $2\frac{7}{8} + 1\frac{5}{6} =$ _____

Solve the problem. Show your work.

6. Maren lives $3\frac{1}{4}$ miles from school. Stormy lives $2\frac{4}{5}$ miles from school. How much farther away from the school does Maren live than Stormy? _____ mile(s)

7. Mark hikes $2\frac{3}{5}$ kilometers to the scenic overlook. He then hikes $1\frac{9}{10}$ kilometers further to the nature center. How many kilometers does Mark hike? _____ kilometers

8. A length of ribbon is $4\frac{2}{9}$ feet long. Millie cuts a piece of ribbon that is $1\frac{5}{6}$ feet long. How long is the remaining piece of ribbon?

 _____ feet

9. Martin buys $2\frac{5}{6}$ pounds of peanuts and $4\frac{3}{4}$ pounds of almonds. How many pounds of nuts does Martin buy?

 _____ pounds

With your child, find two objects around your home whose lengths can be either added or compared. Measure the length of each, using a foot-long ruler, and write the measurements as a mixed number of feet. Have your child add or subtract the two measurements, and then measure the sum or to verify the result. Find two other objects, and repeat as time permits.

Student Practice Book

Lesson 9-9

Additional Practice

Name _____

Review

You can solve problems by adding and subtracting mixed numbers with regrouping.

Carla bought $2\frac{7}{8}$ pounds of cashews and $4\frac{1}{6}$ pounds of raisins for a mixture.

How many pounds is Carla's mixture?	How many more pounds of raisins did Carla buy than cashews?
To solve, find $2\frac{7}{8} + 4\frac{1}{6}$.	To solve, find $4\frac{1}{6} - 2\frac{7}{8}$.
Write the fractions using a common denominator:	Write the fractions using a common denominator:
$2\frac{7}{8} + 4\frac{1}{6} = 2\frac{21}{24} + 4\frac{4}{24}$	$4\frac{4}{24} - 2\frac{21}{24}$
Add the mixed numbers:	Regroup 1 whole as $\frac{24}{24}$:
$2\frac{21}{24} + 4\frac{4}{24} = 6\frac{25}{24}$	$3\frac{28}{24} - 2\frac{21}{24}$
Regroup the fraction:	Subtract the mixed numbers:
$6 + 1\frac{1}{24} = 7\frac{1}{24}$	$3\frac{28}{24} - 2\frac{21}{24} = 1\frac{7}{24}$
Carla's mixture weighed $7\frac{1}{24}$ pounds.	Carla bought $1\frac{7}{24}$ pounds more raisins.

Solve the problem. Show your work.

1. Allison buys $4\frac{1}{3}$ pounds of tomatoes. She uses $2\frac{4}{5}$ pounds of tomatoes to make a salad. How many pounds of tomatoes does Allison have left? _____ pounds

Solve the problem. Show your work.

2. Elijah and Karah went apple picking. Elijah picked $8\frac{5}{6}$ pounds of apples. Karah picked $8\frac{3}{4}$ pounds of apples. How many pounds of apples did Elijah and Karah pick? _____ pounds

3. Herb's water bottle has $12\frac{1}{2}$ ounces of water in it. During a walk, he drinks $8\frac{3}{4}$ ounces of water. How many ounces of water are still in Herb's water bottle? _____ ounces

4. Ashley lives $1\frac{1}{8}$ miles from the school. Eric lives $\frac{3}{4}$ mile from the school. How much closer to the school does Eric live than Ashley?

   _____ mile

5. Jorge plants a flower that is $7\frac{1}{2}$ inches tall. After two weeks, the flower has grown $2\frac{3}{5}$ inches. Now how tall is the flower?

   _____ inches

Math @ Home Activity

With your child, make a set of number cards with the digits 1 through 9. Turn them facedown and select six of the cards. Use them to form two mixed numbers. Add and subtract the mixed numbers. As an additional challenge, try to form the numbers so that the sum is as close to 10 as possible, and the difference is as close to 2 as possible. Then replace the cards and repeat the activity.

Student Practice Book

Lesson **10-1**

Additional Practice

Name _____

Review

You can use a representation to multiply a whole number by a fraction.

Use a representation to find $\frac{3}{5} \times 4$.

Draw 4 wholes. Divide each whole into fifths, or 5 equal parts.	Now shade 3 of the 5 equal parts in each of the 4 wholes, or $\frac{3}{5}$ of each whole.
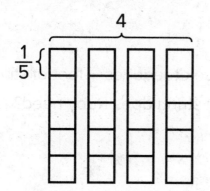	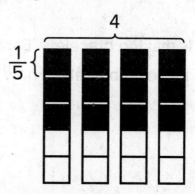
	12 of the fifths are now shaded.

So $\frac{3}{5} \times 4 = \frac{12}{5}$.

Use a representation to find the product.

1. $\frac{2}{3} \times 2 =$ _____

2. $\frac{1}{4} \times 3 =$ _____

Student Practice Book
107

What is the product? Use a representation, if needed.

3. $\frac{3}{5} \times 6 =$ _____

4. $\frac{5}{6} \times 3 =$ _____

5. $\frac{1}{2} \times 5 =$ _____

6. $\frac{2}{3} \times 4 =$ _____

7. Ruby is making bracelets. Each bracelet requires $\frac{4}{5}$ foot of string. If she makes 8 bracelets, how much string does Ruby need?

8. Margie makes 5 recordings. The length of each recording is $\frac{9}{10}$ minute. What is the total length of Margie's recordings?

Write a multiplication equation, such as $\frac{1}{6} \times 24 =$ _____, at the top of a sheet of paper. Ask your child to verbalize what the equation is asking: "There are 24 of something and you are looking for $\frac{1}{6}$ of it." Then have them draw a representation to find the product. Repeat with a different equation.

Student Practice Book

Lesson **10-2**

Additional Practice

Name _____

> **Review**
>
> **You can multiply a fraction and a whole number by multiplying the numerator of the fraction by the whole number, and keeping the denominator the same.**
>
> Johnny walks $\frac{3}{4}$ mile to and from school each day. How many miles did he walk this week if school met all 5 days?
>
> To solve, find $\frac{3}{4} \times 5$.
>
> Multiply the numerator and the whole number: $3 \times 5 = 15$.
>
> This is the numerator. Keep the denominator the same: $\frac{3}{4} \times 5 = \frac{15}{4}$.
>
> Write the answer as a mixed number: $\frac{15}{4} = 3\frac{3}{4}$
>
> Johnny walked $3\frac{3}{4}$ miles to and from school this week.

What is the product?

1. $\frac{5}{6} \times 5 =$ _____

2. $\frac{1}{4} \times 9 =$ _____

3. $\frac{3}{8} \times 6 =$ _____

4. $\frac{2}{3} \times 8 =$ _____

Student Practice Book

5. A bottle of water contains $\frac{3}{4}$ liter of water. How much water do you get when you buy a package of 15 bottles of water?

6. One lap around the track covers $\frac{3}{5}$ mile. Mary walked 8 laps. How many miles did Mary walk?

7. An elephant weighs about 4 tons. A younger elephant weighs about $\frac{5}{8}$ that much. About how many tons does the younger elephant weigh?

8. Margaret buys 20 pounds of flour to use for making treats for the bake sale. She used $\frac{5}{6}$ of the available flour. How many pounds of flour did Margaret use?

With your child, make two stacks of index cards. On the cards in the first stack, write different fractions. On the cards in the second stack, write different whole numbers. Have your child randomly choose a card from each stack. Then have them show you how to find the product of the numbers on the cards. Continue the activity with different pairs of cards.

Student Practice Book

Lesson 10-3
Additional Practice

Name

> **Review**
>
> You can use an area model to multiply a fraction by a fraction. Use each denominator to partition the whole.
>
> Find the product $\frac{1}{4} \times \frac{2}{5}$.
>
> Partition the whole into fifths, and shade 2 of the fifths to show $\frac{2}{5}$. Then partition the shaded fifths into fourths, and shade 1 of the fourths. To make all of the pieces equal size, also partition the unshaded fifths into fourths. The double-shaded part of the area model is the product.
>
>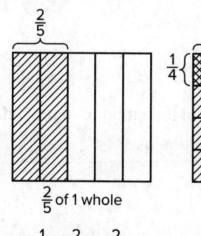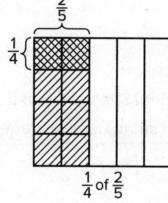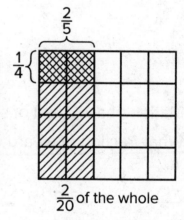
>
> $\frac{2}{5}$ of 1 whole $\qquad$ $\frac{1}{4}$ of $\frac{2}{5}$ $\qquad$ $\frac{2}{20}$ of the whole
>
> So $\frac{1}{4} \times \frac{2}{5} = \frac{2}{20}$.

What is the product? Use the area model.

1. $\frac{3}{4} \times \frac{1}{2} =$ _____

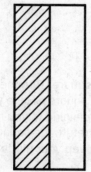

2. $\frac{2}{3} \times \frac{5}{6} =$ _____

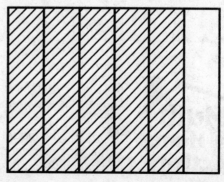

Student Practice Book

What is the product? Use an area model.

3. $\frac{1}{3} \times \frac{4}{7} =$ _____

4. $\frac{2}{5} \times \frac{3}{4} =$ _____

5. $\frac{1}{2} \times \frac{2}{3} =$ _____

6. $\frac{5}{8} \times \frac{1}{3} =$ _____

7. $\frac{2}{5} \times \frac{3}{5} =$ _____

8. $\frac{5}{6} \times \frac{2}{3} =$ _____

9. Marlene has $\frac{3}{5}$ yard of string. She uses $\frac{2}{3}$ of the string for a project. What fraction of a yard of string does Marlene use?

   _____ yard

10. Erica has $\frac{7}{8}$ gallon of water. She needs $\frac{3}{4}$ of the water to water her houseplant. What fraction of a gallon of water does Erica need for her houseplant?

    _____ gallon

Math @ Home Activity

Cut out several paper squares. Write a fraction on each square. Turn the squares facedown and have your child select two of the squares. Have your child use an area model to multiply the two fractions, explaining to you how they made the model and how it shows the result. Then return the squares to the pile and select two more. Repeat as time allows. Keep your models and equations for the next lesson.

Student Practice Book

Lesson 10-4

Additional Practice

Name

Review

You can multiply a fraction by a fraction by multiplying the numerators and multiplying the denominators.

Allen lives $\frac{2}{5}$ mile from the park. He ran $\frac{3}{4}$ of the way to the park, then walked. How far did Allen run?

To solve, find $\frac{3}{4} \times \frac{2}{5}$.

Multiply the denominators of the factors to get the denominator of the product. Multiply the numerators of the factors to get the numerator of the product.

$$\frac{3}{4} \times \frac{2}{5} = \frac{3 \times 2}{4 \times 5} = \frac{6}{20}$$

Allen ran $\frac{6}{20}$ mile on the way to the park.

What is the product?

1. $\frac{2}{3} \times \frac{3}{5} =$ _____

2. $\frac{2}{4} \times \frac{5}{7} =$ _____

3. $\frac{4}{5} \times \frac{3}{7} =$ _____

4. $\frac{6}{11} \times \frac{5}{7} =$ _____

What is the product?

5. $\dfrac{1}{4} \times \dfrac{2}{3} = $ _____

6. $\dfrac{3}{5} \times \dfrac{3}{4} = $ _____

7. $\dfrac{5}{7} \times \dfrac{5}{8} = $ _____

8. $\dfrac{4}{9} \times \dfrac{7}{9} = $ _____

9. $\dfrac{7}{10} \times \dfrac{2}{3} = $ _____

10. $\dfrac{4}{5} \times \dfrac{5}{6} = $ _____

11. A plant is $\dfrac{7}{8}$ foot tall. The plant next to it is $\dfrac{2}{3}$ as tall. How tall is the shorter plant?

_____ foot tall

12. Jessica has $\dfrac{5}{8}$ gallon of water. She drinks $\dfrac{2}{3}$ of it during a walk. How much water did Jessica drink?

_____ gallon

Math @ Home Activity

Use your models and equations from the previous lesson. Have your child multiply the fractions by multiplying the numerators and denominators. Compare the results to the previous results. Ask your child to explain what the product of the denominators represents (the total number of partitioned pieces) and what the product of the numerators represents (the number of double-shaded pieces).

Student Practice Book

Lesson 10-5
Additional Practice

Name _____

> **Review**
>
> **You can find the area of a rectangle with fractional side lengths by tiling the rectangle with unit squares and multiplying the length and width.**
>
> A rectangular garden is $10\frac{1}{2}$ feet long and 6 feet wide. What is the area of the garden?
>
> Find the area of the rectangle.
>
>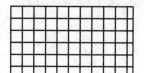
>
> Count the length: $10\frac{1}{2}$ units.
>
> Count the width: 6 units.
>
> Multiply the length and the width to find the area.
>
> Number of whole square units: $6 \times 10 = 60$
>
> Number of half-square units: $6 \times \frac{1}{2} = 3$
>
> Total number of square units: $60 + 3 = 63$
>
> The area of the garden is 63 square units.

What is the area of the rectangle?

1.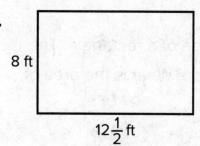

   _____ square feet

2. 2 cm, $20\frac{1}{4}$ cm

   _____ square centimeters

Student Practice Book

What is the area of a rectangle with the given dimensions?

3. 5 inches by $3\frac{1}{2}$ inches

 _____ square inches

4. 8 feet by $4\frac{2}{3}$ feet

 _____ square feet

5. 10 yards by $2\frac{1}{5}$ yards

 _____ square yards

6. 5 meters by $2\frac{3}{4}$ meters

 _____ square meters

7. A ceramic tile is $\frac{3}{4}$ foot wide and $\frac{3}{4}$ foot long. What is the area of the tile?

 _____ square foot

8. Jill's rectangular bedroom is 11 feet long and $9\frac{1}{2}$ feet wide. What is the area of the floor in Jill's bedroom?

 _____ square feet

9. Jake's vegetable garden is in the shape of a rectangle. The garden is 24 feet long and $5\frac{5}{6}$ feet wide. What is the area of Jake's vegetable garden?

 _____ square feet

With your child, look for some rectangles or rectangular objects around your home. Use a ruler or tape measure to measure the dimensions, using fractions of a foot instead of inches. If neither dimension is a whole number, round one of the dimensions to the nearest whole foot. Have your child find the area of the object.

Student Practice Book

Lesson 10-6
Additional Practice

Name _____

Review

You can use an area model to multiply mixed numbers.

Find the product $2\frac{1}{3} \times 3\frac{3}{4}$.

Use an area model.

	3	+ $\frac{3}{4}$
2	6	$\frac{6}{4}$
+ $\frac{1}{3}$	$\frac{3}{3}$	$\frac{3}{12}$

Write each mixed number as a sum. Write the product, or area, in each smaller rectangle.

Add the four partial products: $6 + 1 + \frac{6}{4} + \frac{3}{12}$

Add the whole numbers: $6 + 1 = 7$

Add the fractions: $\frac{6}{4} + \frac{3}{12} = \frac{18}{12} + \frac{3}{12} = \frac{21}{12} = 1\frac{9}{12}$ or $1\frac{3}{4}$

Add the whole numbers and fractions: $7 + 1\frac{3}{4} = 8\frac{3}{4}$

So, $2\frac{1}{3} \times 3\frac{3}{4} = 8\frac{3}{4}$.

What is the product? Complete the area model.

1. $1\frac{2}{3} \times 1\frac{4}{5} = $ _____

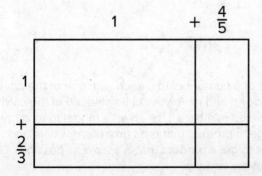

2. $2\frac{5}{6} \times 3\frac{1}{5} = $ _____

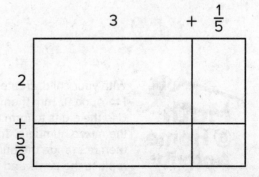

Student Practice Book

What is the product?

3. $2\frac{5}{6} \times 3\frac{1}{4} =$ _____

4. $1\frac{3}{5} \times 8\frac{1}{2} =$ _____

5. $6\frac{1}{3} \times 5\frac{5}{8} =$ _____

6. $2\frac{1}{10} \times 4\frac{3}{4} =$ _____

7. Greg walked $2\frac{2}{5}$ miles yesterday. Today he walked $2\frac{1}{2}$ times as far. How many miles did Greg walk today?

_____ miles

8. Giselle's bedroom wall is $10\frac{3}{4}$ feet long and $8\frac{1}{4}$ feet tall. What is the area of the wall?

_____ square feet

Math @ Home Activity

With your child, create a set of 9 number cards, each with one of the digits 1 through 9. Turn them facedown and have your child select 6 of the cards. Use the digits to form two mixed numbers. Use an area model to multiply the mixed numbers. Try to get a product that is as great as possible and then rearrange the numbers to get a product that is as low as possible. Then replace the cards and repeat the activity

Student Practice Book

Lesson 10-7
Additional Practice

Name _____

> **Review**
>
> **You can multiply mixed numbers by rewriting each mixed number as a fraction greater than 1.**
>
> A rectangular garden is $3\frac{1}{2}$ yards long and $2\frac{1}{4}$ yards wide. What is the area of the garden?
>
> To solve, find the product $3\frac{1}{2} \times 2\frac{1}{4}$.
>
> Write each mixed number as a fraction greater than 1.
>
> $3\frac{1}{2} = \frac{7}{2}$ and $2\frac{1}{4} = \frac{9}{4}$
>
> Multiply the fractions and write the answer as a mixed number.
>
> $\frac{7}{2} \times \frac{9}{4} = \frac{63}{8} = 7\frac{7}{8}$
>
> The area of the garden is $7\frac{7}{8}$ square yards.

What is the product?

1. $3\frac{1}{2} \times 1\frac{1}{2} = $ _____

2. $1\frac{5}{8} \times 2\frac{2}{3} = $ _____

3. $1\frac{3}{5} \times 1\frac{1}{3} = $ _____

4. $4\frac{3}{4} \times 5\frac{2}{5} = $ _____

5. $4\frac{1}{2} \times 2\frac{4}{5} = $ _____

6. $2\frac{1}{3} \times 2\frac{2}{3} = $ _____

7. Walter ran $3\frac{2}{5}$ miles yesterday. Today he ran $1\frac{1}{2}$ times as far. How many miles did Walt run today?

_____ miles

8. Wanda bought a plant that was $1\frac{1}{4}$ inches tall. After two weeks, the plant was $2\frac{1}{2}$ times as tall. How tall was the plant after two weeks?

_____ inches

9. Jodie's backpack weighs $2\frac{3}{8}$ pounds. Jeff's backpack weighs $1\frac{3}{4}$ times as much as Jodie's backpack. How much does Jeff's backpack weigh?

_____ pounds

10. Kyla lives $1\frac{2}{3}$ miles from the park. The library is $2\frac{1}{3}$ times as far from Kyla's house. How far is the library from Kyla's house?

_____ miles

With your child, create a set of 9 number cards, each with one of the digits 1 through 9. Turn them facedown and have your child select 6 of the cards. Use the digits to form two mixed numbers. Write the mixed numbers as fractions greater than 1 and multiply. Try to get a product that is as great as possible and then rearrange the numbers to get a product that is as low as possible. Then replace the cards and repeat the activity.

Lesson 10-8
Additional Practice

Name _____

> **Review**
>
> **You can predict whether a product will be greater or less than one of the factors without performing the multiplication.**
>
> The rosebush is 2 feet tall. The sunflower is $1\frac{1}{2}$ times as tall as the rosebush. The tulip is $\frac{7}{8}$ the height of the rosebush. What is the order of the flowers from shortest to tallest?
>
> | Flower | Height | |
> |---|---|---|
> | Rosebush | 2 feet | |
> | Sunflower | $2 \times 1\frac{1}{2}$; taller than rosebush | When multiplying by a factor greater than 1, such as $1\frac{1}{2}$, the answer will be greater than the factor. |
> | Tulip | $2 \times \frac{7}{8}$; shorter than rosebush | When multiplying by a factor less than 1, such as $\frac{7}{8}$, the answer will be less than the factor. |
>
> From shortest to tallest, the order of the flowers is tulip, rosebush, and sunflower.

1. Which expressions have a value greater than 38? Choose all that apply.

    A. $38 \times \frac{3}{2}$ B. $38 \times \frac{4}{3}$

    C. $38 \times \frac{5}{8}$ D. $38 \times \frac{5}{2}$

    E. $38 \times \frac{10}{6}$ F. $38 \times \frac{3}{7}$

Student Practice Book

Circle the lesser number.

2. $7 \times \frac{9}{10}$ or $7 \times \frac{5}{2}$

3. $15 \times \frac{6}{5}$ or $15 \times \frac{5}{6}$

4. $\frac{1}{3} \times 2$ or $\frac{7}{5} \times 2$

5. $\frac{15}{4} \times 20$ or $\frac{12}{15} \times 20$

Write a fraction that makes each sentence true.

6. $\frac{1}{2} \times$ _____ $< \frac{1}{2}$

7. $\frac{8}{13} \times$ _____ $> \frac{8}{13}$

8. $\frac{15}{6} \times$ _____ $> \frac{15}{6}$

9. $\frac{7}{9} \times$ _____ $< \frac{7}{9}$

10. On Monday, Willie ran $3\frac{1}{2}$ miles. On Wednesday, he ran $\frac{3}{5}$ that distance. On Friday, he ran $1\frac{3}{5}$ times that distance. What is the order of the days from shortest run to longest run?

With your child, practice predicting whether a product will be greater than or less than a certain factor. For example, if the distance from your home to school is 2 miles, and your child's friend lives $\frac{1}{2}$ the distance from the school, have your child determine who lives farther from the school.

Student Practice Book

Lesson 10-9
Additional Practice

Name _____

Review

You can solve problems involving multiplication of fractions and mixed numbers.

Mrs. Adler has a poster that is $14\frac{1}{2}$ inches long and $8\frac{1}{2}$ inches wide. What is the area of the poster?

Use a representation.	Rewrite mixed numbers as fractions greater than 1.
(area model: 14 + ½ by 8 + ½; 112, 4, 7, ¼) $112 + 7 + 4 + \frac{1}{4} = 123\frac{1}{4}$	$14\frac{1}{2} \times 8\frac{1}{2} =$ $\frac{29}{2} \times \frac{17}{2} =$ $\frac{493}{4} = 123\frac{1}{4}$

The area of the poster is $123\frac{1}{4}$ square inches.

1. Bella is painting a sign that is $4\frac{3}{4}$ feet long and $3\frac{1}{2}$ feet wide. What is the area of the sign?

   _____ square feet

2. Ariana lives $\frac{5}{8}$ mile from school. Fred lives $\frac{3}{4}$ the distance from school as Ariana. How far does Fred live from the school?

   _____ mile

3. There are 42 flowers in a vase. Of the flowers, $\frac{1}{2}$ are roses. Of the roses, $\frac{3}{7}$ are white. How many white roses are in the vase?

   _____ white roses

4. A photograph of a skyline is enlarged to be $2\frac{1}{4}$ feet tall and 6 feet long. What is the area of the enlarged photograph?

   _____ square feet

5. At the basketball game, $\frac{9}{10}$ of the people are cheering for the home team. Of those people, $\frac{3}{4}$ of them are wearing team shirts. What fraction of the people at the game are cheering for the home team wearing a team shirt?

   _____ of the people

6. A family is building a rectangular patio in the backyard. The patio is to be $5\frac{3}{4}$ yards long and $3\frac{1}{4}$ yards wide. What is the area of the patio?

   _____ square yards

7. In a fifth-grade class, $\frac{3}{4}$ of the students play an instrument. Of those students, $\frac{2}{5}$ of them play the piano. What fraction of the students in the class play the piano?

   _____ of the students

Math @ Home Activity

With your child, identify situations around your home where multiplication of fractions and mixed numbers would be necessary. For example, if the length of a piece of fabric is $\frac{3}{4}$ foot long and the width of the fabric is $2\frac{1}{2}$ feet long, students would multiply to find the area of the piece of fabric. Ask your child to show how to solve the problem in two different ways. Have them explain how each strategy yields the same product.

Student Practice Book

Lesson 11-1
Additional Practice

Name

> **Review**
>
> **You can interpret a fraction as another way to write a division expression.**
>
> Tynisha cuts a wooden board that is 4 feet long into 3 equal sections. What is the length of each piece of wood?
>
> To solve, find $4 \div 3$.
>
> Draw 4 wholes and divide each into 3 equal pieces.
>
>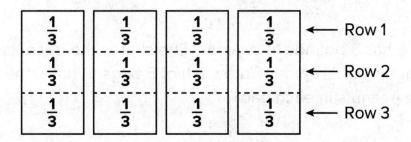
>
> The total for each row is the quotient. Since each row contains 4 one-thirds of a foot, the length of each piece is $\frac{4}{3}$ or $1\frac{1}{3}$ feet long.
>
> $4 \div 3 = \frac{4}{3}$ or $1\frac{1}{3}$

What division expression is represented by the fraction?

1. $\frac{12}{5}$

2. $\frac{1}{4}$

3. $\frac{3}{8}$

4. $\frac{9}{2}$

What fraction is represented by the division expression?

5. $7 \div 6 =$ _____

6. $12 \div 3 =$ _____

7. $2 \div 5 =$ _____

8. $6 \div 8 =$ _____

9. $10 \div 10 =$ _____

10. $1 \div 4 =$ _____

11. Giselle has 3 pounds of peanuts. She shares the peanuts by putting an equal amount into each of 5 bags. What is the weight of the peanuts in each bag?

12. Juan walks 8 miles. He divides the walk into 3 equal parts so he knows when to stop for water. How far does Juan walk between stops?

13. Aubrey draws a line that is 34 centimeters long. She divides the line into 6 equal parts. How long is each part of the line?

Provide opportunities for your child to explore how fractions and division are related. For example, ask him or her to determine how much each person would receive if a given amount, such as 2 pounds of granola, was divided equally among each person in your family. Have your child write a division expression and a fraction for the situation.

Student Practice Book

Lesson 11-2

Additional Practice

Name _____

Review

You can determine whether the quotient should be written with a remainder or as a mixed number.

A pitcher holds 42 fluid ounces of lemonade. Helene pours an equal amount into each of 5 glasses until the pitcher is empty. How much lemonade does Helene pour into each glass?

To solve, find $42 \div 5$.

With a remainder, $42 \div 5 = 8$ R2.

As a mixed number, $42 \div 5 = 8\frac{2}{5}$.

Since fractional parts of fluid ounces can be poured, write the answer as a mixed number.

Helene poured $8\frac{2}{5}$ fluid ounces into each glass.

How would you write the quotient for the problem?

1. Callie walked a certain number of miles last week. She walked the same number of miles each day. How many miles did she walk each day?

    A. as a mixed number

    B. with a remainder

    C. either way is appropriate

2. Debbie made some bracelets. She gave the same number of bracelets to her friends. How many bracelets did she give to each friend?

    A. as a mixed number

    B. with a remainder

    C. either way is appropriate

3. A 10-kilometer race is divided into 3 equal sections. How long is each section of the race?

4. A teacher orders a box of 100 pencils to give to the students. Each of the 18 students receives the same number of pencils. How many pencils does each student get?

5. A fence is 40 yards long. Fence posts are placed so that there are 6 equal sections. How far apart are the fence posts?

6. A grocer has 50 peaches to sell. He packages them in groups of 3. How many packages does the grocer make?

Math @ Home Activity

Provide opportunities for your child to explore how division might be represented with a remainder or with a mixed number. For example, if you make a 64-fluid ounce pitcher of a drink, and you have 10 plastic cups, how many fluid ounces will be in each cup if all cups have the same amount of the drink?

Student Practice Book

Lesson **11-3**
Additional Practice

Name

Review

You can use a representation to find the quotient of a whole number divided by a unit fraction.

Rosanna has 5 large pieces of fabric. To make a quilt, she needs to cut each large piece of fabric into 5 pieces or $\frac{1}{5}$s. How many smaller pieces of fabric will she have?

To solve, find $5 \div \frac{1}{5}$.

Use a representation to find the quotient. Draw 5 wholes. Divide each whole into $\frac{1}{5}$s.

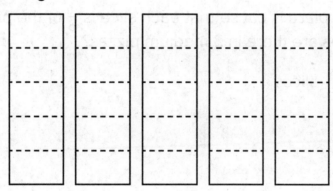

There are 25 pieces that are $\frac{1}{5}$ of a whole.

Rosanna will have 25 smaller pieces of fabric.

What is the quotient? Use a representation to solve.

1. $3 \div \frac{1}{8} = $ _____

2. $8 \div \frac{1}{5} = $ _____

3. $4 \div \frac{1}{4} = $ _____

4. $2 \div \frac{1}{6} = $ _____

5. Carl has a board that is 4 feet long. He makes shelves that are $\frac{1}{2}$ foot long. How many shelves can he cut from the board?

_____ shelves

6. A baker has 8 pounds of flour. Each cake needs $\frac{1}{3}$ pound of flour. How many cakes can be made with the available flour?

_____ cakes

7. A medium pizza is cut so that each slice is $\frac{1}{6}$ of the pizza. How many slices are there in 3 medium pizzas?

_____ slices

8. A caterer makes 6 pans of fruit salad. Each serving is to be $\frac{1}{10}$ the size of the pan. How many cups of fruit salad can be served?

_____ cups

Give your child several sheets of paper. First, have your child fold one sheet of paper in half. Ask them how many sections were created. Then have your child calculate how many sections there would be if a given number, such as 5, 6, or 7 sheets of paper, was folded the same way. Repeat the activity with different numbers of sections.

Student Practice Book

Lesson **11-4**

Additional Practice

Name _____

> **Review**
>
> **You can use multiplication to find the quotient of a whole number divided by a unit fraction.**
>
> A baker has a 4-pound bag of flour. A recipe uses $\frac{1}{5}$ pound of flour. How many times can the baker make the recipe?
>
> To solve, find $4 \div \frac{1}{5}$.
>
> There are 5 $\frac{1}{5}$-pounds in each pound of flour.
>
> | 1 | $\frac{1}{5}$ | $\frac{1}{5}$ | $\frac{1}{5}$ | $\frac{1}{5}$ | $\frac{1}{5}$ |
> |---|---|---|---|---|---|
> | 2 | $\frac{1}{5}$ | $\frac{1}{5}$ | $\frac{1}{5}$ | $\frac{1}{5}$ | $\frac{1}{5}$ |
> | 3 | $\frac{1}{5}$ | $\frac{1}{5}$ | $\frac{1}{5}$ | $\frac{1}{5}$ | $\frac{1}{5}$ |
> | 4 | $\frac{1}{5}$ | $\frac{1}{5}$ | $\frac{1}{5}$ | $\frac{1}{5}$ | $\frac{1}{5}$ |
>
> So there are $4 \times 5 = 20$ $\frac{1}{5}$-pounds in 4 pounds.
>
> The baker can make the recipe 20 times.

What is the quotient?

1. $9 \div \frac{1}{8} =$ _____

2. $11 \div \frac{1}{5} =$ _____

3. $6 \div \frac{1}{4} =$ _____

4. $5 \div \frac{1}{6} =$ _____

5. $3 \div \frac{1}{2} =$ _____

6. $8 \div \frac{1}{3} =$ _____

7. A large bag of granola weighs 6 pounds. If smaller bags are made that each contain $\frac{1}{8}$ pound of granola, how many smaller bags can be filled?

_____ smaller bags

8. A watermelon weighs 7 pounds. Slices are cut so that each piece weighs $\frac{1}{4}$ pound. How many slices are cut from the watermelon?

_____ slices

9. Rosita's garden is 8 yards long. She plants a row of flowers in her garden. She plants each flower $\frac{1}{3}$ yard apart. How many flowers does Rosita plant?

_____ flowers

10. Winnie has 9 apples. She cuts each apple so that each slice is $\frac{1}{5}$ of the apple. How many apple slices does Winnie have?

_____ slices

Math @ Home Activity

Write the numbers 2 through 12 on separate index cards. Next write the unit fractions with denominators 2 through 10 on separate index cards. Place the whole-number cards and the fraction cards in two separate piles, and have your child randomly select a card from each pile. Then have them divide the whole number by the fraction. Repeat the activity, putting the cards back into their respective piles each time.

Student Practice Book

Lesson 11-5
Additional Practice

Name _____

Review

You can use a representation to find the quotient of a unit fraction divided by a whole number.

Belinda uses $\frac{1}{2}$ of her flower garden for roses. She plants 4 rosebushes, giving each an equal amount of the garden. What fraction of Belinda's flower garden will be used for each rosebush? To solve, find $\frac{1}{2} \div 4$.

Use a representation to find the quotient.

Draw $\frac{1}{2}$ of one whole to show the part of the garden for the roses.	Divide the $\frac{1}{2}$ into 4 equal parts for each rosebush.
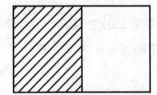	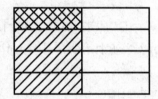

Each rosebush will use $\frac{1}{8}$ of the flower garden.

What is the quotient? Use a representation to solve.

1. $\frac{1}{6} \div 4 =$ _____

2. $\frac{1}{4} \div 2 =$ _____

3. $\frac{1}{9} \div 5 =$ _____

4. $\frac{1}{5} \div 3 =$ _____

Student Practice Book
133

What is the quotient?

5. $\frac{1}{3} \div 8 = $ _____

6. $\frac{1}{7} \div 6 = $ _____

7. $\frac{1}{10} \div 2 = $ _____

8. $\frac{1}{9} \div 4 = $ _____

9. In 3 minutes, Javier can walk $\frac{1}{6}$ mile. How far does Javier walk in 1 minute?

_____ mile

10. A baker has $\frac{1}{2}$ pound of flour. From this amount, the baker can make 5 cakes. How much flour does the baker use to make each cake?

_____ pound

11. A swimmer swims 5 lengths of the pool to swim $\frac{1}{4}$ kilometer. What fraction of a kilometer is each length of the pool?

_____ kilometer

Math @ Home Activity

Set out measuring cups and measuring spoons that represent unit fractions, $\frac{1}{3}$ cup or $\frac{1}{4}$ teaspoon. Have your child practice dividing each unit fraction into 2, 3, or 4 smaller, equal amounts. Use other measuring cups or spoons to verify the results, if possible.

Lesson **11-6**

Additional Practice

Name _____

> **Review**
>
> **You can use multiplication to find the quotient of a unit fraction divided by a whole number.**
>
> Mr. Torres has $\frac{1}{3}$ of a large container of glue to divide equally among 2 smaller containers. How much of the glue in the large container will be put into each small container?
>
> To solve, find $\frac{1}{3} \div 2$.
>
> Use multiplication to find the quotient.
>
> Dividing by 2 is the same as multiplying by $\frac{1}{2}$.
>
> $\frac{1}{3} \div 2 = \frac{1}{3} \times \frac{1}{2} = \frac{1}{6}$
>
> Each small container can hold $\frac{1}{6}$ of the glue from the larger container.

What is the quotient?

1. $\frac{1}{5} \div 7 = $ _____

2. $\frac{1}{8} \div 3 = $ _____

3. $\frac{1}{6} \div 9 = $ _____

4. $\frac{1}{3} \div 5 = $ _____

5. $\frac{1}{4} \div 6 = $ _____

6. $\frac{1}{9} \div 2 = $ _____

7. Greta draws a line that is $\frac{1}{2}$ foot long. She divides the line into 4 equal sections. What is the length of each section?

_____ foot

8. Joseph lives $\frac{1}{5}$ mile from school. He can walk to school in 5 minutes. How far does Joseph walk each minute?

_____ mile

9. Karlie still has $\frac{1}{3}$ of her book left to read. She plans to read the same amount each of the next 5 days. How much of the book does Karlie plan to read each day?

_____ of the book

10. A pitcher of lemonade is $\frac{1}{4}$ full. Remy pours the lemonade equally into 3 cups. What fraction of a full pitcher of lemonade gets poured into each cup?

_____

Math @ Home Activity

With your child, look for situations around your home where fractional amounts are present. For example, if $\frac{1}{4}$ of a meal is left over, ask your child to determine how much of the original meal each person in your family will receive if the leftovers are shared equally. Use a unit fraction for the amount of leftovers. Look for and solve other examples.

Student Practice Book

Lesson 11-7

Additional Practice

Name

> **Review**
>
> **You can use strategies you know to help you solve problems involving division.**
>
> A sandwich shop uses $\frac{1}{4}$ pound of lunch meat on its sandwiches. Yesterday, the sandwich shop used 20 pounds of lunch meat. How many sandwiches were served yesterday?
>
> To solve, find $20 \div \frac{1}{4}$.
>
> There are four $\frac{1}{4}$s in each whole.
>
> So, $20 \times 4 = 80$.
>
> The sandwich shop served 80 sandwiches yesterday.

1. Deanne covers $\frac{1}{3}$ of her notebook cover with 5 stickers. Each sticker is the same size. What part of the entire notebook cover does each sticker cover?

   _____

2. Marvin uses a mix and some water to make 54 fluid ounces of fruit punch. He pours an equal amount into 8 glasses for himself and seven friends. How much fruit punch does each person get?

   _____ fluid ounces

Student Practice Book

3. A baker has 10 pounds of flour on hand. Each batch of cookies needs $\frac{1}{2}$ pound of flour. How many batches of cookies can the baker make using the available flour?

_____ batches

4. Maxine has 2 pounds of raisins. She places an equal amount into each of 15 snack bags. How many pounds of raisins are in each snack bag?

_____ pound

5. Andrea has 50 perennials to plant. She plants the flowers in 6 equal rows, using as many flowers as possible. How many perennials are in each row? How many are left unplanted?

_____ perennials in each row;

_____ perennials left unplanted

6. Matthew has $\frac{1}{3}$ pound of trail mix. He eats all of it in 4 equal servings during his hike. How much trail mix does Matthew eat in one serving?

_____ pound

Math @ Home Activity

With your child, look for situations around your home where your child can practice solving problems involving division. For example, if there are 3 apples left, and 5 people each want some, how much does each person get if they share equally? Look for and solve other examples that have been studied in this unit.

Student Practice Book

Lesson 12-1
Additional Practice

Name _____

Review

You can use multiplication or division to convert customary units of measurement and units of time.

To convert from a larger measure to a smaller measure, multiply because there will be more of the smaller unit.

To convert from a smaller measure to a larger measure, divide because there will be fewer of the larger unit.

Write equivalent measures in yards and in inches for 6 feet.

Use the equivalent measures:

1 yard = 3 feet Divide since the conversion is from feet to yards: $6 \div 3 = 2$.	1 foot = 12 inches Multiply since the conversion is from feet to inches: $6 \times 12 = 72$.

So 6 feet is equivalent to 2 yards and to 72 inches.

Which operation do you use for the conversion?

1. hours to minutes

2. gallons to quarts

3. ounces to pounds

4. inches to feet

Student Practice Book

What is the equivalent measure?

5. 120 min = _____ h

6. 3 lb = _____ oz

7. 48 mo = _____ yr

8. 10 ft = _____ in.

9. 2 gal = _____ qt

10. $\frac{2}{3}$ hr = _____ min

11. A football team has to advance 10 yards to get a first down. How many feet is this? _____ feet

12. The running time for a movie is 180 minutes. How many hours long is the movie? _____ hours

13. Betty bought 2 gallons of milk. Jane bought 6 quarts of milk. Who bought more? How much more?

Provide opportunities for your child to convert customary units. For example, measure the dimensions of a window in feet and have your child show how to convert the measures to inches. Or if your child has 60 minutes until bedtime, ask them to tell how many hours it is until bedtime.

Student Practice Book

Lesson 12-2
Additional Practice

Name _____

> **Review**
>
> You can use multiplication or division to convert metric units of mass, length, or capacity.
>
> To convert from a larger measure to a smaller measure, multiply because there will be more of the smaller unit.
>
> To convert from a smaller measure to a larger measure, divide because there will be fewer of the larger unit.
>
> Write equivalent measures in meters and in millimeters for 45 centimeters.
>
> Use the equivalent measures:
>
> | 1 meter = 100 centimeters | 1 centimeter = 10 millimeters |
> |---|---|
> | Divide since the conversion is from centimeters to meters: | Multiply since the conversion is from centimeters to millimeters: |
> | 45 ÷ 100 = 0.45 | 45 × 10 = 450 |
>
> So 45 centimeters is equivalent to 0.45 meter and to 450 millimeters.

Which operation do you use for the conversion?

1. kilograms to grams

2. milliliters to liters

3. meters to kilometers

4. grams to milligrams

Student Practice Book

What is the equivalent measure?

5. 3 m = _____ cm

6. 5.2 L = _____ mL

7. 240 g = _____ kg

8. 1,200 m = _____ km

9. 500 mg = _____ g

10. 40 mL = _____ L

11. Phyllis ran in a 5-kilometer race to help raise money for the school. How many meters long is the race? _____ meters

12. Jenny's water bottle holds 1.3 liters of water. How many milliliters of water does the water bottle hold? _____ milliliters

13. Joe's dog has a mass of 6.52 kilograms. How many grams is the mass of Joe's dog? _____ grams

Provide opportunities for your child to convert metric units. For example, scales can measure in pounds but also in kilograms or grams. Have your child convert a number of kilograms to a number of grams, or grams to kilograms. Containers often show measures in fluid ounces, but also in liters or milliliters. Measurements of length can be made in meters or centimeters—have your child explain how to convert a measure to the other unit.

Student Practice Book

Lesson 12-3
Additional Practice

Name

> **Review**
>
> You can convert units of measurement to help you solve problems that have multiple steps.
>
> Kathryn has a new spool of ribbon that holds a total length of 2.5 meters of ribbon. She uses 225 centimeters of ribbon to wrap some gift boxes. How much ribbon does Kathryn have left?
>
> To solve, find 2.5 meters − 225 centimeters.
>
> First, convert 2.5 meters to an equivalent measure in centimeters: 2.5 × 100 = 250, so 2.5 m = 250 cm.
>
> Then subtract: 250 cm − 225 cm = 25 cm.
>
> Kathryn has 25 cm of ribbon left.

1. Zach has a pitcher that holds 1.5 L of lemonade. Each cup holds 280 mL of lemonade. He pours out 5 glasses for himself and 4 friends. How much lemonade will be left in the pitcher?

2. A bag of apples weighs 3 pounds. Each apple weighs 6 ounces. How many apples are in the bag? _____ apples

Student Practice Book

3. A hiking trail is 3.2 km long. From the start of the trail, the bridge is 1.4 km along the trail. Once at the bridge, a waterfall is 900 m farther. How far is it from the waterfall to the end of the trail?

4. A deli uses 4 ounces of meat on each of its sandwiches. How many sandwiches can be made from 5 pounds of meat?
   _____ sandwiches

5. A bush is 3 feet 4 inches tall. Herb trims the bush so that the bush is now $\frac{3}{4}$ the height it was. How tall is the bush now?

Look for situations in which two of the same types of measurements occur, such as millimeters and centimeters, or feet and inches. With your child, create a problem situation that involves the two measures. Have your child explain how to solve the problem by converting one of the measures to an equivalent measure using the other unit.

Lesson **12-4**

Additional Practice

Name _____

> **Review**
>
> You can create a line plot from a set of data and use it to make observations about the data.
>
> The times, in hours, that Kaylee practices the piano are shown. Which time or times occurs most often?
>
> $\frac{1}{4}, 1, \frac{1}{2}, 1, 1\frac{1}{2}, \frac{3}{4}, 1$
>
> To solve, make a line plot of the data.
>
> Make a number line showing all of the possible times. Use an X to mark one occurrence.
>
> **Piano Practice Times (hours)**
>
>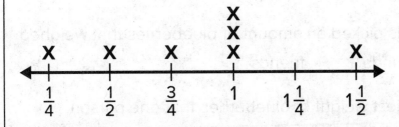
>
> Since there are 3 X only above 1, the time that occurs most often is 1 hour.

Use the line plot above for questions 1 and 2.

1. How many times did Kaylee practice for 1 hour or more?
   _____ times

2. How many times did Kayle practice for exactly $1\frac{1}{4}$ hours? Explain.

Student Practice Book

A group of friends picked blueberries. The weights, in pounds, of the amounts of blueberries each person picked are listed.

$\frac{3}{4}, 1\frac{1}{4}, 1\frac{3}{4}, 2, \frac{3}{4}, 1, \frac{3}{4}, 1\frac{1}{4}, 1, \frac{3}{4}, 2, 1$

3. Make a line plot of the data.

Weight of Blueberries (pounds)

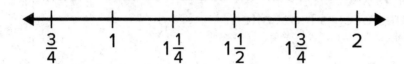

4. How many friends were in the group? _____ friends

5. Which weight or weights of blueberries were picked most often?

   _____ pound(s)

6. How many friends picked an amount of blueberries that weighed more than $1\frac{1}{2}$ pounds? _____ friends

7. What is the heaviest weight of blueberries that one person picked? _____ pounds

Give your child 10 objects with lengths that are within 1 inch of each other. Have him or her measure the objects, record the lengths, and then create a line plot for the data. Ensure that your child includes key components of a line plot, such as a title, labeled tick marks, and an X for each data value. Have your child make observations about the data from the line plot.

Lesson **12-5**
Additional Practice

Name _____

Review

You can solve problems by interpreting information given in line plots and then performing operations.

Brett makes a line plot showing the amounts of water, in cups, that he drinks throughout the day. What is the total amount of water, in cups, that Brett drinks during the day?

Amounts of Water (cups)

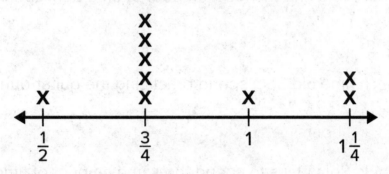

To solve, add the amounts represented by each X. Use multiplication to help.

$$\frac{1}{2} + (5 \times \frac{3}{4}) + 1 + (2 \times 1\frac{1}{4}) = \frac{1}{2} + \frac{15}{4} + 1 + \frac{10}{4}$$

Using a common denominator, $\frac{2}{4} + \frac{15}{4} + \frac{4}{4} + \frac{10}{4} = \frac{31}{4}$

As a mixed number, $\frac{31}{4} = 7\frac{3}{4}$.

Brett drank $7\frac{3}{4}$ cups of water throughout the day.

Use the line plot above for questions 1 and 2.

1. What is the difference between the greatest amount of water Brett drank and the least? _____ cup

2. How many times did Brett get a drink of water during the day? _____ times

Sem made a line plot to show the time, in hours, that he spends practicing the guitar each day for a week.

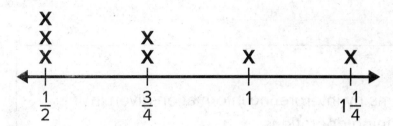

3. For how many days did Sem record his time spent practicing? _____ days

4. What is the difference between the greatest amount of time Sem spent practicing and the least amount of time Sem spent practicing? _____ hour

5. How much time did Sem spend practicing the guitar during the week? _____ hours

6. Next week, Sem plans to spend the same amount of time practicing, but plans to spend an equal amount of time each day. How much time should Sem spend practicing each day?

_____ hour(s)

Look for opportunities around your home where your child can measure the lengths or weights of similar objects that have fractional measurements, such as the heights of some books on a bookshelf. Have them record the measurement data and then create a line plot to represent the data. Ask your child questions about the data on the line plot that requires them to perform calculations with the fractional numbers to find the answer.

Lesson **13-1**
Additional Practice

Name _____

Review

You can represent a point on a coordinate plane using an ordered pair.

What ordered pair represents the point on the coordinate plane where A is located?

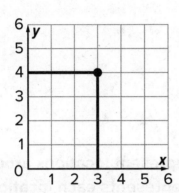

Counting along the x-axis from the origin, A is at 3. So 3 is the x-coordinate of A.

Counting along the y-axis from the origin, A is at 4. So 4 is the y-coordinate of A.

An ordered pair is of the form (x-coordinate, y-coordinate).

The ordered pair (3, 4) represents point A.

What is the ordered pair that represents the point on the coordinate plane?

1. R _____
2. S _____
3. T _____
4. U _____
5. V _____
6. W _____

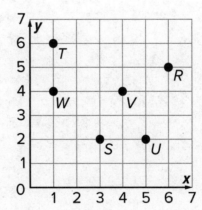

What is the ordered pair that represents the point on the coordinate plane?

7. A (5, 4)
8. B (2, 2)
9. C (2, 5)
10. D (3, 3)
11. E (3, 1)
12. F (1, 3)
13. G (4, 5)
14. H (5, 2)
15. I (2, 4)
16. J (1, 5)

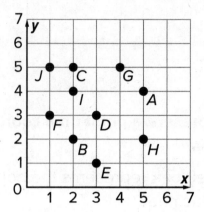

Conrad uses a coordinate plane to represent locations around his town. What is the ordered pair that represents each location?

17. Home (5, 3)
18. School (4, 1)
19. Grocery Store (2, 5)
20. Library (1, 1)

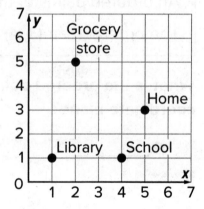

Create a coordinate plane that includes labels for the *x*- and *y*-axes. Give your child a marker and have them mark several points on the plane. Then have them give each point a different label. Work with your child to identify the ordered pair that represents each point on the coordinate plane. Ask your child to explain how they determined the ordered pairs.

Student Practice Book

Lesson 13-2
Additional Practice

Name _____

Review

You can plot a point on a coordinate plane if you are given an ordered pair.

How do you plot point A at (4, 3) on the coordinate plane?

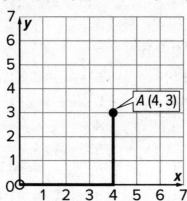

The x-coordinate is 4 and the y-coordinate is 3. From the origin, move 4 units to the right, along the x-axis. Then move up 3 units, along the y-axis.

Label point A at (4, 3).

Plot the point for each ordered pair. Label with the given letter.

1. A (5, 3)
2. B (4, 1)
3. C (2, 5)
4. D (1, 2)
5. E (4, 5)
6. F (5, 2)
7. G (1, 4)
8. H (3, 4)
9. I (2, 1)
10. J (3, 3)

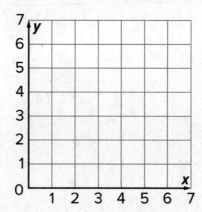

Student Practice Book
151

Plot and label the point for each of the following positions.

11. Catcher (0, 0)
12. Second Base (5, 3)
13. Pitcher (3, 3)
14. Shortstop (2, 4)
15. First Base (5, 1)
16. Third Base (1, 4)

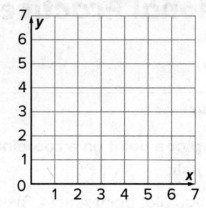

17. Monica wants to plot the point (5, 2) on a coordinate grid to represent the position of her mailbox. Did she plot the point correctly? Explain.

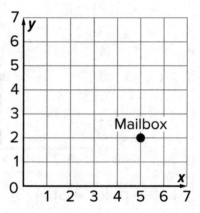

Using 10 index cards, write the name of a location on the front of each card and an ordered pair on the back of each card. Give your child the cards and a coordinate grid. Have them plot each location using a different color. Then randomly pick a card and have them explain how they plotted the point.

Student Practice Book

Lesson **13-3**

Additional Practice

Name

Review

You can interpret points on a coordinate grid to help you understand real-world problems.

The graph shows the number of pages Connie read over 7 days. On which day did Connie read the greatest number of pages?

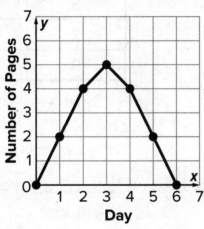

The highest point along the y-axis, 5, shows the greatest number of pages read on one day. The x-coordinate of that point on the graph is 3.

Connie read the greatest number of pages on Day 3.

Use the graph above for questions 1–3.

1. How many pages did Connie read on Day 2? _____ pages

2. On which day(s) did Connie read 2 pages? Day(s)

3. What does the point (6, 0) mean?

Will flies a drone in his yard. An app on his phone records the time the drone is in the air and its height. The table shows the results.

Drone Height	
Time (s)	Height (ft)
0	6
1	12
2	18
3	24
4	24
5	18
6	12
7	6
8	0

4. Plot the points on the coordinate grid to represent the height of the drone for each number of seconds that it is in the air. Then connect the points.

5. From what height does the drone take off? _____ feet

6. How high was the drone at 3 seconds? _____ feet

7. What does the point (7, 6) mean?

8. What was the drone doing between 3 seconds and 4 seconds after it took off?

Create a table of values that could represent a situation your child is familiar with. Situations can include the number of minutes spent practicing an instrument or the number of minutes spent reading a book. Have your child plot the points and then connect the points with line segments. Point to different points on the graph, and ask your child to explain what the point means in relationship to the given context.

Student Practice Book

Lesson 13-4
Additional Practice

Name

Review

You can classify triangles as scalene, isosceles, or equilateral based on the number of sides that have equal length.

Scalene triangles have no sides the same length.

Isosceles triangles have at least two sides the same length.

Equilateral triangles have all three sides the same length.

The tick marks show sides that have equal length.

Scalene	Isosceles	Equilateral

Classify the triangle using the terms *scalene*, *isosceles*, **and** *equilateral*.

1.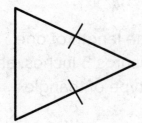

   _____

2.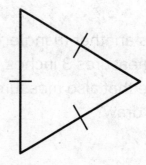

   _____

Student Practice Book
155

Classify the triangle using the terms *scalene*, *isosceles*, and *equilateral*.

3.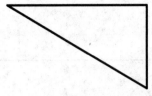

4.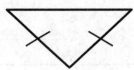

_____   _____

5.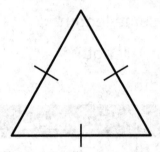

6.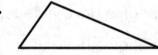

_____   _____

7. Ezra draws a triangle that has the length of one side equal to 10 centimeters, one side that has a length that is less than 10 centimeters, and one side that has a length that is greater than 10 centimeters. What type of triangle does Ezra draw?

8. Ezra draws another triangle. This triangle has the length of one side that measures 3 inches, one side that measures 5 inches, and a third side that also measures 5 inches. What type of triangle does Ezra draw?

With your child, be on the lookout for different triangles that you may see in your everyday experiences. For example, you might notice that a Yield traffic sign is in the shape of an isosceles triangle. Look for other examples and classify the triangles according to the number of sides that are the same length.

Lesson 13-5
Additional Practice

Name _____

Review

You can identify quadrilaterals by their properties. The arrows indicate that the sides are parallel. The tick marks indicate that the sides are the same length.

A trapezoid is a quadrilateral with exactly one pair of parallel sides.	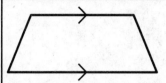
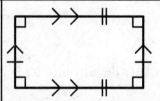	A parallelogram is a quadrilateral with two pairs of parallel sides and two pairs of sides that are the same length.
A rectangle is a parallelogram with four right angles.	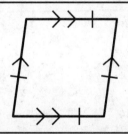
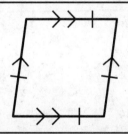	A rhombus is a parallelogram with four sides of equal length.
A square is a parallelogram with four sides of equal length and four right angles.	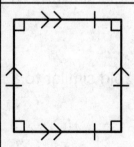

Student Practice Book

1. What are some properties that a quadrilateral may have? List as many as you can.

2. How is a rhombus similar to a square?

3. How is a rectangle similar to a parallelogram?

4. How is a trapezoid different from a parallelogram?

5. How is a rectangle similar to a square?

6. How is a trapezoid similar to a rectangle?

Have yourself and your child create riddles using the descriptions of the quadrilaterals in this lesson. For example, "I have four right angles, my opposite sides are parallel, and my opposite sides are the same length. What am I?" (rectangle). Then exchange riddles and try to determine the type of quadrilateral. Discuss any differences or inaccuracies in the riddles.

Student Practice Book

Lesson **13-6**
Additional Practice

Name

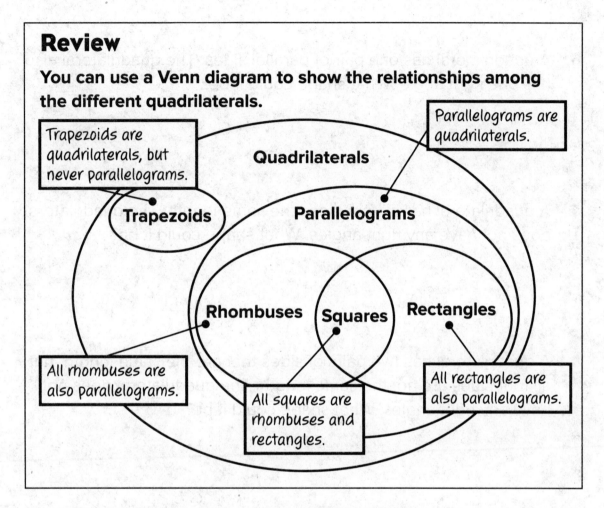

Decide whether the statement is TRUE or FALSE.

1. All rectangles are parallelograms. _____

2. All rhombuses are squares. _____

3. All squares are rectangles. _____

4. A trapezoid can never be a parallelogram. _____

Student Practice Book

5. A quadrilateral has two pairs of sides that are parallel. The quadrilateral also has four right angles. What shape could it be?

6. A quadrilateral has one pair of parallel sides. The quadrilateral also has one right angle. What shape could it be?

7. A quadrilateral has all four sides the same length. The quadrilateral does not have any right angles. What shape could it be?

8. A quadrilateral has two pairs of sides that are the same length, but all four sides are not the same length. The quadrilateral does not have any right angles. What shape could it be?

9. Jesse draws a quadrilateral so that two sides measure 8 inches and the other two sides measure 5 inches. The shape has all right angles. What shape could it be?

Have yourself and your child create riddles using the descriptions of the quadrilaterals in this lesson. For example, "I have four right angles, my opposite sides are parallel, and my opposite sides are the same length. What am I?" (rectangle). Then exchange riddles and try to determine the type of quadrilateral. Discuss any differences or inaccuracies in the riddles.

Lesson **14-1**
Additional Practice

Name _____

> **Review**
>
> **You can use numbers, operation symbols, such as +, −, ×, and ÷, and grouping symbols, such as ( ), to write numerical expressions.**
>
> Charissa cuts up some oranges into 30 slices. She gives 2 slices to her sister and then divides the remaining slices equally among 4 friends. Write a numerical expression to represent how many orange slices each friend will get.
>
> First, subtract 2 from 30: 30 − 2
>
> Then divide the result by 4: (30 − 2) ÷ 4
>
> Each friend will receive the number of orange slices represented by the numerical expression (30 − 2) ÷ 4.

What numerical expression represents the description?

1. Multiply 6 and 7. Then add 5.

2. Subtract 2 from 8. Multiply the difference by 3.

3. Add 4 and 7. Then divide 44 by the sum.

4. Divide 18 by 3. Multiply 4 and 5. Then add the quotient and the product.

Student Practice Book

5. Each bag of nuts and raisins contains 6 ounces of nuts and 4 ounces of raisins. Write a numerical expression to represent how many ounces of nuts and raisins are needed to make 20 bags of nuts and raisins.

6. Kristin cuts several apples into 46 slices. She gives 6 to her brother and then divides the remaining apple slices equally among her 5 friends. Write a numerical expression to represent how many apple slices each of her friends will get.

7. Greta plants her flowers in 5 rows of 8 plants, and then plants the remaining 3 flowers in another row. Write a numerical expression to represent how many flowers Greta planted.

8. A set of pens contains pens that write with different colors of ink: 4 blue, 3 black, 2 red, and 1 purple. Write a numerical expression to represent how many pens a teacher will have if 12 sets of pens are ordered.

Identify situations in your everyday experiences where packages of different items may be purchased. Have your child identify the number of different items in each package. Then ask them to write a numerical expression to determine the number of specific items that will be in a certain number of packages.

Lesson **14-2**

Additional Practice

Name _____

> **Review**
>
> **You can understand the relationship between numbers by interpreting the numerical expressions.**
>
> What is the same about the numerical expressions $(20 \div 4) + 6$ and $20 \div (4 + 6)$? What is different?
>
> The numerical expressions are the same in that both use the same numbers, 20, 4, and 6, the same operations, division and addition, and both use grouping symbols.
>
> The grouping symbols, however, make the numerical expressions different because different numbers are grouped together.
>
> The numerical expression $(20 \div 4) + 6$ means to divide 20 by 4, then add 6. The numerical expression $20 \div (4 + 6)$ means to divide 20 by the sum of 4 and 6.

Write the description for each numerical expression.

1. $(11 \times 9) + 5$

2. $11 \times (9 + 5)$

3. $20 - (12 \div 4)$

4. $(20 - 12) \div 4$

Compare the expressions using >, <, or =. Explain your reasoning.

5. 60 ÷ 10 ◯ (60 ÷ 10) + 7

6. 40 × 6.5 ◯ (40 − 8) × 6.5

7. 5 × (4 + 3$\frac{1}{2}$) ◯ (5 × 4) + (5 × 3$\frac{1}{2}$)

8. (20 × 15) − 42 ◯ 20 × 15

Tell how the value of the first numerical expression compares to the value of the second numerical expression.

9. 512 + 259 and (512 + 259) × 3

10. (28 × 43) + 12 and 28 × 43

11. (36 ÷ 4) − 3 and 36 ÷ 4

Math @ Home Activity

Write the four mathematical symbols (+, −, ×, ÷) on separate index cards. Write random numbers on 16 other index cards. Create a numerical expression using two symbol cards and three numbers cards. Have your child explain a situation that could be represented by the expression. Then have them create an expression for which you will determine a situation.

Student Practice Book

Lesson **14-3**
Additional Practice

Name _____

> **Review**
>
> **You can evaluate a numerical expression using the order of operations.**
>
> Evaluate the numerical expression $4 + 6 \times (10 - 3)$.
>
> Evaluate within grouping symbols first.
>
> $4 + 6 \times (10 - 3) = 4 + 6 \times 7$
>
> Perform any multiplication or division, in order from left to right.
>
> $4 + 6 \times 7 = 4 + 42$
>
> Perform any addition or subtraction, in order from left to right.
>
> $4 + 42 = 46$
>
> The numerical expression $4 + 6 \times (10 - 3)$ evaluates to be 46.

Which operation will you perform first to evaluate the expression? Explain your reasoning.

1. $32 + 7 \times (8 - 3)$

2. $42 + 10 \div 5 - 2$

3. $8 \div 2 \times 4 + 6$

4. $10 - 6 + 100 \times 4$

Evaluate the numerical expression.

5. $10 - 5 + 2$

6. $6 + 12 \div 6$

7. $(3 + 4) \times 3$

8. $15 - (2 + 7) + 1$

9. $24 \div 2 \times 6 + 1$

10. $8 \div (2 \times 2) + 1$

11. $2 \times 9 - 8 + 1$

12. $14 - (6 + 7) + 4$

13. $42 \div 6 - 3 + 4 \times 5$

14. $4 + 36 \div (6 \div 3 + 4) \times 5$

15. $5 \times (12 - 2 \times 5) + 36 \div (10 - 6 + 2)$

Write a 3- or 4-step numerical expression at the top of a sheet of paper. Give your child four different color pencils. Assign a color to each of the steps. Have your child evaluate the expression, using the correct color to show progression from one step to the next. Repeat the activity with a different expression.

Student Practice Book

Lesson 14-4

Additional Practice

Name _____

> ## Review
>
> **You can generate numerical patterns using rules and identify a relationship between corresponding terms in two numerical patterns.**
>
> Erika and Leo are picking apples. The first minute they each pick 0 apples. Then each minute after, Erika adds 4 apples to her basket and Leo adds 8 apples to his basket. When Erika has 16 apples in her basket, how many apples will Leo have in his basket?
>
> Number of apples in Erika's basket each minute: 0, 4, 8, 12, **16**, 20
>
> Number of apples in Leo's basket each minute: 0, 8, 16, 24, **32**, 40
>
> When Erika has 16 apples in her basket, Leo will have 32 apples in his basket.

Refer to the numeric patterns for Erika and Leo above.

1. What is the rule for the number of apples in Erika's basket?

2. What is the rule for the number of apples in Leo's basket?

3. What is the relationship between corresponding terms in the two patterns?

4. When Leo has 48 apples in his basket, how many apples will Erika have in her basket? _____ apples

Student Practice Book

167

5. Write the first six terms of the numerical pattern that starts at 0 and follows the rule Add 3.

6. Write the first six terms of the numerical pattern that starts at 0 and follows the rule Add 6.

7. Compare the numerical patterns. What is the relationship between corresponding terms in the two patterns?

Rodney counts the value of his pennies. Diane counts the value of her nickels. They both start with 0 coins worth 0 cents.

Value of Rodney's pennies: 0, 1, 2, 3, 4, 5, 6

Value of Diane's nickels: 0, 5, 10, 15, 20, 25, 30

8. What is the rule for Rodney's pattern?

9. What is the rule for Diane's pattern?

10. What are the next three numbers in Rodney's pattern?

11. What are the next three numbers in Diane's pattern?

12. What is the relationship between corresponding terms in the two patterns?

13. When Diane has 40 cents, what will be the value of Rodney's coins? _____ cents

14. When Rodney has 10 cents, what will be the value of Diane's coins? _____ cents

Write two numerical patterns horizontally on a sheet of paper. Ask your child to identify the rule for each pattern, and have them show you how to write the next 3 terms for each pattern. Then have them explain how to find the relationship between corresponding terms in the two patterns, and predict the corresponding number in one pattern given a term in the other pattern.

Student Practice Book

Lesson 14-5

Additional Practice

Name _____

> **Review**
>
> You can organize numerical patterns in a table to help you identify and describe relationships between corresponding terms and use this relationship to determine unknown terms.
>
> Pattern A starts at 0 and adds 2 to each term.
>
> Pattern B starts at 0 and adds 8 to each term.
>
> What is the corresponding term in Pattern B when the term in Pattern A is 14?
>
> Make a table to show the first 5 terms in each pattern:
>
> | Pattern A | 0 | 2 | 4 | 6 | 8 |
> |-----------|---|---|----|----|----|
> | Pattern B | 0 | 8 | 16 | 24 | 32 |
>
> Notice that the terms in Pattern B are 4 times the corresponding terms in Pattern A.
>
> When 14 is the term in Pattern A, the corresponding term in Pattern B is $14 \times 4 = 56$.

Refer to the numeric patterns A and B above.

1. When the term in Pattern A is 22, what will be the corresponding term in Pattern B?

2. When the term in Pattern B is 48, what will be the corresponding term in Pattern A?

3. When the term in Pattern B is 200, what will be the corresponding term in Pattern A?

4. When the term in Pattern A is 100, what will be the corresponding term in Pattern B?

Use the patterns for problems 5–8.

Pattern A: Starts at 0 and adds 3 to each term.

Pattern B: Starts at 0 and adds 9 to each term.

5. Complete the table that shows the first six terms of the numerical patterns.

Pattern A						
Pattern B						

6. What is the relationship between the terms in Pattern B and the corresponding terms in Pattern A?

7. When the term in Pattern A is 21, what will be the corresponding term in Pattern B?

8. When the term in Pattern B is 90, what will be the corresponding term in Pattern A?

9. A recipe requires 2 ounces of flour for every 6 ounces of water. A baker uses 12 ounces of flour. How many ounces of water should the baker use? _____ ounces

10. A restaurant uses 3 eggs in every omelet served. How many omelets were served if 24 eggs were used? _____ omelets

Write two rules for numerical patterns on a sheet of paper. Ask your child to make a table showing the first 6 terms for each pattern. Have them explain the relationship between the corresponding terms in the two patterns. Then have them predict the corresponding number in one pattern given a term in the other pattern.

Student Practice Book

Lesson 14-6
Additional Practice

Name _____

Review

You can plot corresponding terms from numerical patterns as ordered pairs on a coordinate plane and use the data to solve problems.

The table shows the number of hours driven on a car trip and the number of miles traveled. The ordered pairs are graphed on the coordinate plane.

Number of hours	Number of miles	Ordered pair
1	40	(1, 40)
2	80	(2, 80)
3	120	(3, 120)
4	160	(4, 160)
5	200	(5, 200)

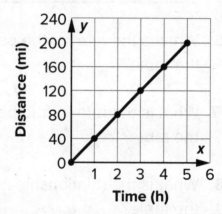

How many hours will it take to travel 240 miles?

Extend the line of the graph. When the line reaches 240 miles, along the y-axis, the x-coordinate of the point will be 6.

It will take 6 hours to travel 240 miles.

Refer to the data above for problems 1–4.

1. How many miles were traveled after 3 hours? _____ miles

2. How many hours did it take to travel 80 miles? _____ hours

3. What is the relationship between the number of hours spent traveling and the number of miles traveled?

4. After 10 hours, how many miles will be traveled? _____ miles

Student Practice Book

5. Fran earns $8.00 per hour at her job. Complete the table showing the number of hours Fran works and the amount of money she earns. Then graph the ordered pairs.

Number of Hours	Amount Earned ($)	Ordered pair
0	0	(0, 0)

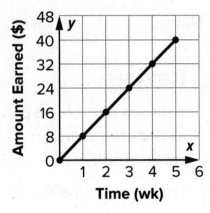

6. What is the rule for the pattern in the Number of Hours column in the table?

7. What is the rule for the pattern in the Amount Earned column in the table?

8. What is the relationship between the corresponding terms in the table?

9. How much money does Fran earn for working 4 hours? $ _____

10. How many hours does Fran have to work in order to earn $40? _____ hours

11. How much money will Fran earn for working 8 hours? $ _____

12. How many hours does Fran have to work in order to earn $80? _____ hours

Math @ Home Activity

Gather a handful of nickels. Have your child determine the number of cents when there are 0, 1, 2, 3... nickels and record the data in a table. Then have them create the ordered pairs represented by the relationship. Work together to create a graph of the relationship. Then ask your child questions about the information before having them determine the rule showing the relationship between the number of nickels and the number of cents.

Student Practice Book